TRAVIS STONE

RISE

OF THE

MACHINES

SECRET WEAPONS, SECRET WARS, & SECRET AGENDAS

RISE OF THE MACHINES
Secret Weapons, Secret Wars,
& Secret Agendas

ISBN-10: 1508586284

ISBN-13: 978-1508586289

First Print Edition: February 2015

Made in the United States of America

A Product of ICU Press

Author's Note

Two broad types of 'secret' military technology exist;
conventional - and un-conventional.

Unmanned Aerial Vehicles, UAVs; Unmanned
Aerial Systems, UASs; Unmanned Combat Aerial
Vehicles, UCAVs; Remotely Piloted Aircraft, RPA or
'*drones*', fall into the conventional field, and although
many are not secret, all are essentially, in today's
technological paradigm, quite simple machines.

When one considers the technology, drones
should have been operating since the 1980's . . . oh,
that's right - they were - just in *secret*.

It is rumored that the Nazis even flew drones
late in the Second World War.

The fact that the F-117A Nighthawk, a piloted
stealth strike aircraft, and the Predator and Global
Hawk drones were never known about (in their times of
priority) is a testament to the Military/ Industrial
security classification system at the level of Top-Secret.
The system works.

But now that we know of the next generation
stealth drones like the RQ-180 Sentinel, what *now* is the
real top-secret technology?

Author's Note

Some 'in the know' talk of a massive *50 year* technology gap between what we see and what actually exists in the shadow worlds of 'above top-secret' covert strikes, aerial assassinations, and borderline surveillance. This term 'technology gap' however, is a broad expression; weaponizable technology spans a myriad of technology groups, each with their own 'technology gap'. Un-conventional technology like electromagnetic scalar and beam weapons sit at different levels to the 'conventional' drones. Un-conventional technology lives above Top-Secret, or to be precise, outside of military classification systems, and likely outside of military knowledge.

Drones however, have a specific role or roles. Drones *could* now be built to replace obsolete Interceptor, Air-superiority, Fighter-bomber, and Close Air Support Aircraft (CAS) like the F-22 Raptor, F-16 Falcon, F-18E Super Hornet, & the old A-10 Warthog. There is also nothing stopping the 'drone-ization' of an actual F-22 - or even a Boeing 777-300ER for that matter.

The drones *we* know (because we see them on TV or read about them in articles) are Military and CIA systems used in the Middle East and North Africa to strike specific targets, or *"To take-out 'high-value' Al Qaeda targets."*

But the reality - and we must face this without denial - is that another generation of 'drones' will now be able to occupy near-space.

Are drones off planet?

With (at least) a 50 year 'technology gap' in the type of machines and technology we can see on the internet and those which exist in the secret U.S. black-black programs, we can posit that space drones have been operating for quite some time - and it is certainly time for us to stop pretending otherwise.

This idea however, throws up hard questions like: do other nations operate drones 'off planet'? And: is there a secret (but human) 'space war', or a 'cold space war' currently ongoing in our solar system?

What do we really know of our reality?

This might sound crazy, but when all is considered, we would be foolish to think otherwise.

In the following pages, I will take the Rise of the Machines a step further. I will examine drones and their tactics in regards to Mideast objectives. I will examine covert wars and assess their underlying motives. If past behavior is the best indicator of future behavior, I will examine a brief history of drones, airpower, mindsets, and motives concerning the need for concealing the reality of secret wars from the people who pay for them - the public.

But behind such wars lies a complex web of ever evolving secret sciences, and secret technologies of horrific power, masked by very convincing deception -

The real 'secret wars' are far more frightening than most could ever know.

Travis Stone, 17/11/2014, Oxford

CONTENTS

Contents

Secret Space Wars

Secret Secret Wars

Af-Pak: A War Within A War Within A War Within A War

The well publicized 'secret war' in Afghanistan and Pakistan is of course, not secret.

The CIA's architects of deception have primed the media and fed them specific information and footage obtained by their many secret surveillance systems - the intended result is what I call a 'Deception Availability Cascade'. An Availability Cascade is where the general public is led to incorrect conclusions by means of Media sensationalism; a Deception Cascade being where this effect is engineered by an intelligence group.

Aside from putting the 'terror' in terrorism, this CIA targeted media blitz seeds the perception that the combined drone & Special Forces operations in Af-Pak are 'secret', when they are clearly not.[1]

There is a simple reason for this simple deception: now any covert, borderline, or illegal drone

---

1. Quinion. (2009). "Af-Pak". World Wide Words. Af-Pak combines the nations of Afghanistan & Pakistan into one 'war-zone', for U.S. military purposes.

strikes and illegitimate assassinations in the greater Mideast can be connected to this marginal, but publicly accepted 'secret war'. However, this 'public secret war' masks an actual secret war - which is necessary to achieve the Middle Eastern corporate goals of our Military/ Industrial juggernaut. Of course we know this, and to many it seems not that much of a big deal, but nonetheless, these twists of deception underlie illegalities that are never publicized - instead what we see is highly emotive terror footage in a media cascade designed for three distinct purposes:

1) Generate a disproportionate fear of Islamic terrorism; whilst simultaneously feeding terror cells and allowing them prolific freedom to inflict emotive and gruesome acts of terror.

2) Generate support for Western military action in the Mideast, and;

3) Provide cover for corrupt or secret CIA operations in the Mideast.

When the reality is veiled by lies underpinned by expensive and detailed deception operations, it clearly suggests activity that would fail to gain public (or Congressional) support - or in the case of the Mideast, activity that in many cases constitutes U.S. war crimes.

We the public however, find ourselves with no option but to support the War on Terror; but it is nonetheless a war which masks a Military/ Corporate money-go-round, and it is a war engineered by deception planners to appear as a necessary act of self-defense.

But does this media driven CIA publication of 'secret' Middle Eastern wars serve another purpose - a more sinister purpose? Is such a ruse a diversion - news media entertainment to keep us all occupied whilst *other* events play-out - like a magician luring your eye away from the trick?

Whatever the case, there is clearly more to the secret Af-Pak drone war than meets the eye.

What *is* the true nature of the secret-secret war?

To 'go there', an examination of the Military/ Industrial command and security classification structure is necessary.

Theorized U.S Military/Industrial Security Structure In Place During The 1990s

Majestic
Cosmic
Luna
Ultra
Stellar
Astral
Cosmos
Triad
Orbit
ZD-27

Top-Secret Crypto
Levels 1 → 28 (Compartmentalized)
(President of United States sits at the default level of Crypto 17, but is subject to change.)

Unacknowledged Special Access Programs (USAP)
Special Compartmented Information (SCI)
Top-Secret
Secret
Confidential
Restricted
Un-Classified

Regardless of any name changes or alterations to the system (the system above crypto *will* have changed) the necessary points can still be made:

1) Military commanders (Generals, Admirals & below) will not be privy to higher-level (corporate) objectives - they are simply given instructions to carryout, likely believing in false threats and therefore, false goals. In-fact, this could be true of each and every compartmentalized level.

2) Corporate interests sit above governmental interests.

3) Some select government officials will be cleared into the upper echelons of MIC security compartmentalization.

4) Military Intelligence operates within the Top-Secret-Crypto compartment.

5) Some select military officials will be cleared into Crypto.

6) The CIA operates above military bands; however, different rank bands within the CIA operate in very different realities due to the information, internal deception, and goals that are fed to them. This concept will become clear later.

6) Corporate, CIA, Government, Science, Contractors, and Military are separated, and then

compartmentalized. This not only controls what people in the system know, it controls what they believe.

If this chart's accuracy is even remotely close to reality, then it also suggests a group (the Majestic Group?) is the overseeing power of the global financial/ military-industrial system, and the U.S. President, and the U.S. military, are simply tools under Majestic control.

Vannevar Bush was a founding member of Majestic.[2]

In reference to the Military Industrial Security System, President Dwight D. Eisenhower, said in January of 1961.

> "The *potential* for the disastrous rise of misplaced power exists, and will persist."[3] (italics added)

This is an interesting statement for Eisenhower to make for several reasons: As Commander of Allied Expeditionary Forces in Europe during WWII, he was already a tool of that 'potential' system.

Did he not know?

Eisenhower maintained constant contact with Winston Churchill and Franklin D. Roosevelt, the lynchpins between corporate influence and military action, so he must've known the extent of corporate

---

2. Stone. ATime for Deception
3. Ibid

influence *already* dictating U.S. military policy in 1944.

Corporate interests have controlled the government, which in turn has controlled the military since the passenger ship *Lusitania* was loaded with what Winston Churchill referred to as "live bait", and was then steamed un-escorted into the German declared 'war-zone' where she was sunk by U-20 at 2.10pm on May 7, 1915, killing 1257.[4]

The point in relation to Mideast drone wars of course, is that the U.S. military thinks its drone war in Af-Pak is related to Counter-Terrorism, when the greater percentage is dictated by corporate objectives.

And these 'corporate objectives' are the real 'secret war', and the real reason for occupying Af-Pak and Iraq.

The CIA's 'secret war' operates separately and at a higher level, but under the same principles.

If you have not read my previous book *A Blurred Reality: Isis, Power, Terror, & Deception,* I suggest doing so will clear the fog surrounding Middle Eastern secret wars, and put the need for the drone in a framed picture. While *A Blurred Reality* carries the evidence and builds accurately to reveal Mideast deception, I will list the pertinent points here, although they may appear out of context without their supporting facts:

---

4. Stone. A Time for Deception

1) Within the Middle East, the U.S. functions as an Empire.

2) The CIA functions as the Middle East's Gestapo.

3) They do so, not just to keep terrorists from our boarders, but to control the flow, availability, and therefore the price of Mideast oil - and to stop others from doing the same.

4) These Mideast terrorist groups are armed, and to some extent, controlled by the CIA and Saudi Arabian financers. This is clear - it is the extent of this 'assistance' that remains blurred.

5) These terror cells are real; their members believe in Jihad; they hate America.

6) The first terror cells were a creation of the CIA's 1979 arming and nursing of the Mujahideen. The CIA armed and funded the Mujahideen so as to lure the USSR into a ten year war, causing the collapse of the Soviet Military, and the collapse of the Soviet Union in 1991, by fighting to 'the last Afghan boy' to win.

7) Today the CIA 'allows' IS to exist so as to:

> a) Heighten our fear of terror attacks, or a second 911 on U.S. soil, justifying Iraq's occupation under the guise of self defense, and;

> b) To topple non-compliant dictators inside the Mideast oil ring, maintaining a buffer of plausible deniability for illegal, ruthless control of the old

British oil Empire - this charade is transparent to many, including Vladimir Putin, and many American politicians.

c) To trigger destructive Middle Eastern civil wars; because the Mideast cannot be allowed to become technologically advanced or educated as this would threaten U.S. control. A continued and periodic 'regression' via CIA triggered civil wars ensures U.S./corporate control over the Mideast going forward.

d) This of course is denied, and the prevailing illusion of terror (although very real) is exacerbated and exaggerated; the purpose of the prevailing illusion is to make us support a 'just war' of 'self-defense', which when all the camouflage is removed, is really a semi-permanent occupation.

8) 9/11, whether orchestrated entirely by the Bush cartel; allowed to happen; or whether it happened because of incompetence, has achieved the patriotism and public support needed to occupy and pulverize Iraq for the purposes of the above reasons.

9) The $68 billion per year Intelligence Community uses ignorance and incompetence as its excuse for 'not seeing [insert terrorist act] coming' - for what other excuse is there?

10) The façade of deception is crumbling. The charade is transparent.[5]

We (The U.S.) are in Af-Pak/ Iraq for the purposes of both physically controlling our Empire, and controlling the threat of terrorism on U.S. soil - the threat from these Mideast countries however, is:

1) Inflated to increase our fear and the justification for maintaining continual presence there.

2) Currently a relatively low risk (statistically) when compared to car crashes and illness, but a media driven 'deception availability cascade' incorrectly inflates the risk to fearful levels.

3) A very *real* risk exists nonetheless because our 'terminator' style treatment and ruthless control of oil revenue creates a terrorist mindset when combined with a history of extreme Islamic brainwashing, currently getting a steroid injection via IS. The risk posed by terrorism is real but low; however in reference to IS specifically, the risk is localized within the Mideast.

4) The CIA (under Stansfield Turner and Zbigniew Brzezinski) actually built and created 'Islamic schools' in Afghanistan from 1979 and into the 1980s. These schools taught vulnerable Afghan boys an

---

5. Stone. A Blurred Reality. pg80

extreme form of Islamic Jihad. The CIA did this because they knew Afghanis *would* fight the USSR to the last boy, but only if brainwashed into an extreme Jihadist mentality.

5) If it is true, and the CIA currently has people inside IS, and is arming, funding and controlling their spread - whilst they could stop them but instead choose to "*Slowly degrade* and push them back." Then we *will* see an increase in gruesome terror attacks designed for social media - the spread of fear making it impossible for a terrified western public to denounce the United States' Mideast occupation - just as the fear of Soviet Nuclear attack melted public resistance to the pointless pre-emptive invasions of Korea and Vietnam - and the bogus paper invasions of Cuba, drawn up by the JCS in 1962, and to be justified by false-flag acts of self-inflicted terrorism, as out laid in the 1962 *Operation Northwoods* briefing document. (Attached as Annex I)

Justified by a *Northwoods* influenced approach, the Korean and Vietnam Wars severed no benefit to the American people; they did however create massive corporate benefit for both the emerging MIC and U.S. oil companies.

But is our current-to-future reality heading in an even more horrific direction?

Judge, Jury, & Terminator

Apex Predator: The Grim Reaper

Produced by General Atomics Aeronautical Systems, the MQ-1 Predator became the first UAV to take up a strike role for both the USAF and the CIA.

Official sources place first operational production in 1995, but many believe the Predator was combat operational as early as 1991 over the former Yugoslavia.

The MQ-1 carries two AGM-114 Hellfire missiles, and with a flight time of 20-22 hours, can fly out 400 nautical miles, loiter for 14 hours, and still return safely to its base in Afghanistan or Djibouti.

The Predator has been the mainstay for both drone presence and CIA assassination strikes in the Mideast since 2001, but is set to be phased out in favor of the MQ-9 Reaper, essentially a heaver version of the Predator. The Reaper is powered by a single turboprop engine, possessing a reliability and power that the Predator's piston engine cannot match. The Reaper's extra power lets it carry a staggering 15-times more weapons than the Predator. A Reaper drone costs 17 million USD (2014), which may seem steep, but is a drop in the ocean when compared to the F-22 Raptor's $150 million dollar price tag.

The difference of course, is the Raptor, and its replacement the F-35, are complete air-superiority stealth fighters; their role is very different to that of the lightweight terminator drones - the F-35 will match any known aircraft, and more to the point, will fight and win in a real war.

I say 'real war' because in the Mideast there exists none of the following:

1) Air Defense/ SAM or Surface to Air Missile sites, or;

2) Interceptor or air-to-air capable aircraft.

3) An enemy force with anything better than RPGs.

Due to the lack of air defenses, Reapers and Predators conduct Mideast specific ISR missions in total safety. (ISR = Intelligence, Surveillance, & Reconnaissance.)

As advanced satellite and 'other' airborne surveillance systems are available over the Mideast, it suggests that drone surveillance is specific to 'see-and-shoot' type operations - AKA assassination.

Justified under the War on Terror's self-defense policy, targets are 'taken out' with no respect for evidence, or a trial - only the grim verdict of behavioral recognition software. Mistakes are made - lots of them.

Innocent people are killed.

But do we care?

Does the end justify the means?

As we know, Reaper pilots mostly fly these search & destroy missions from mainland America, 7000 odd miles away from the actual machine.

This fact gives the public a feeling of comfort - knowing that none of our pilots are in real danger - but short of mechanical failure, a piloted fighter-bomber faces no danger over the Middle East either.

The difference is that drones provide:

a) Ultra long loiter time over target-area (14-20 hours)

b) The ability to swap pilots at any time.

c) A reduction of the need for 'good pilots', as drone flights employ specialist takeoff and landing crews.

d) The ability for a CIA 'Intelligence Officer' to stand 'over-the-shoulder' of a drone pilot, giving specific instructions.

e) The ability of a CIA operative to actually fly a drone assassination mission, and then hand off to a landing crew.

f) The ability to do so from secret or undisclosed locations, creating opportunities for 'missions that never happened' - or 'off the books' assassinations.

g) Is it relevant to point out that during the 1940s, Hitler's Gestapo would've loved nothing more than this ability: the ability to fly, loiter, and strike in total deniability?

The significance of these points will become clear later.

To the humans scurrying like rats on the ground, the Predators and Reapers are nothing less than faceless robotic terminators, who could at any moment, turn, assess them as criminals, and destroy their lives in a second - without though or emotion, and without justification or questions ever being asked - and certainly without conviction in a court of law.

Clearly, if terrorists are killed it makes the world safer - but not all kills are just, many are mistakes, and more often than not, civilians pay with their lives.

The CIA and its brutal new form of 'sometimes accurate' justice, has become Judge, Jury, & Terminator.

In terms of drone killing, clearly ethical and moral dilemmas exist. These dilemmas dominate virtually every 'drone debate'. The question of whether U.S./ CIA counter-terror action is legal, is not however, the goal of this work. My goal is to discover what is *not* being *said*.

What are the deceptions?

With deception wrapping every word and image that we see and hear coming from the Pentagon, it stands to reason (under the doctrine of deception) that we are being manipulated to think of the Mideast drone wars in a certain way.

We will not know the full story.

That part we have to investigate, study, and work out for ourselves.

The most basic rules of deception as the Pentagon operates, are of course:

1) To appear weak where you are strong; and vice-versa;

2) To appear strong where you are weakest.

3) To cover real operational motives with false motives, and;

4) To cover 'Above-Top-secret' technology with 'Top-secret' but visible technology. This point is a staple of Cold War deception, and critical to 'national security'.

With this in mind, one may start to unstitch the cover of Mideast deception. A point to also note is: CIA deception may fool the Western public and the Intelligence apparatus of a few countries, but it does not fool Russia's. Russia, to a large extent, remains quiet and compliant with Western objectives, which contradicts what we are told (via the media) about Russian intentions, raising several points and questions:

1) Is Vladimir Putin in league with U.S./ CIA covert/ corporate operations within the Mideast?

2) If so, what hold does the Military Industrial conglomerate have over Russia? Are corporate links in place?

3) Does this hold revolve around the flow and price of Russian oil exports, or other trade or commodity related controls?

4) If not, why does Putin not expose CIA deception going back to the Mujahideen's funding and creation, the Vietnam oil war, and the nursing of al-Qaeda by both Saudi and U.S. interests?

5) If not aimed at Russia's Intelligence apparatus, Mideast deception must be designed for us - the Western public (who also pays the bills). As we will see though, deception is also aimed at other bands of the Military/ Industrial Security Command, in the bizarre dance of deception known as 'cover'.

All deception plans designed for Mideast operations are fed to us (ranging from the taxpaying public, to high-level public servants, to CIA station Chiefs, to corporate CEOs) by the media. The TV tells us what to believe, and this means two things:

1) Mideast deception is designed for the Western public, not an enemy or enemy Intelligence apparatus.

2) That it is very easy to manipulate our thinking because 95% of people in the West have a TV; and news articles are both professional, convincing, and easily manipulated.

The point is: Mideast deception does all of these things:

1) Appears weak in areas of strength.

2) Covers operational motives with false ones.

3) Covers super-secret technology with visible 'top-secret' technology, i.e.: are drones covered by scalar weapons?

4) Uses the media to deceive the public.

5) Must deceive the public because if allowed to understand true corporate motives and control, the system of the U.S. Middle Easter Empire could fail, and;

6) Because we might start seeing where the Western future is really heading.

A Few Obvious Anomalies

In the Mideast, drone strikes have killed thousands of 'high value al-Qaeda targets'. Although the exact number is classified, the Washington post published the figure of 3000 al-Qaeda kills in a 2013 article.

Targets are supposedly chosen from a compilation of names called 'the kill list', approved outside of Afghanistan and North West Pakistan only by President Obama (2014); inside Af-Pak, no Presidential approval is required for drone assassinations.

The Anomalies are:

1) Despite the thousands of reported drone strikes over 13 years, the kill list remains full.

2) Although denied, as many civilians as al-Qaeda targets have been killed by drones, driving new waves of anti-western sentiment.

3) Military hunter/ killer drone operators such as the USAF and the Joint Special Operations Command (JSOC) will have absolutely no idea how many civilians

have been hit (or targeted) by secret CIA drone strikes. The CIA will feel no obligation to inform them.

4) The Military operators will believe in their cause, because they are fed the same deception matrix as the public. They are proud and believe they are helping to eliminate al-Qaeda - while the CIA, operating under entirely different, corporate controlled motives, feeds them deception and conducts missions which appear to support IS and al-Qaeda.

3) There is no end in sight; because of their 'ineffectiveness' drones may have to operate in the Mideast for decades.

4) Paradoxically, despite their 'ineffectiveness' at eliminating Al Qaeda and now IS leaders, the CIA considers drones its most effective tool in the Mideast; their funding, production, and prioritization growing exponentially. Regarding this point: if drones can't stop Mideast terror, what do they provide the CIA that makes them so desirable?

Is it because:

a) Drone strikes stir up anti-western mindsets so as to *maintain* IS and Al Qaeda in order to justify the United States' Middle East occupation?

b) Drones terminate anyone hindering, or counter to CIA goals - 'off the books'?

c) Despite possessing airpower and ground tactics that could wipe-out IS's relatively weak force, drones create the illusion of a concerted effort to remove IS - but one that will never succeed?

d) The CIA wants to insert 'secret' drone operations into our mindsets as 'the norm', grooming us to accept a future of drone operations in the United States?

The Rise of the Fourth Reich

Is a killer-drone surveillance state the future of U.S. law enforcement? Will we loose our freedom of speech, and freedom to protest, in a gradual, Nazi style change of culture to the new Gestapo - the CIA?

If they don't care who they kill in Pakistan or Yemen or Africa, why will they care about 'collateral damage' on U.S. soil?

If they fund and fuel terrorism and then fail to control it, why will they care about you and your family?

The great irony will be: when the drones come here, our freedom is over - *freedom,* the single justification for drone killings in the first place!

Don't for a second think that our CIA operators won't behave this way. The Nazis duped their military and their general population into mass murder, world war, and total loss.

This didn't happen because German people are inferior thinkers or stupid. It happened through deception - clever, prolonged, well planed, incremental deception, employing the following key factors:

1) Making the population afraid; afraid of communists; afraid of their neighboring armies' imminent attack.

2) Telling the German population they were under attack by the Polish - when it was Germany who had attacked first.

3) Using these false-flag attacks (self inflicted terrorism) to create fear and generate patriotism.

4) Using 'self-defense' and 'Homeland Security' to fuel patriotism, when Nazi paramilitary units were the real aggressors.

5) Using this mass patriotism to invade peaceful countries in 'pre-emptive' wars of conquest justified by 'self defense' - the people happy with this because they were scared.

6) Identifying the population's narcissists and psychopaths and promoting them into positions of leadership within the Nazi party, the SS, and the Gestapo.

7) Making the changes to fascism incrementally: the Nazis didn't start gassing Jews and launching blitzkrieg war over Europe in one step - hundreds of incremental steps are used to effect change in any environ - each small step into the worlds of Jew gassing or robotic drone assassination prepares you for the

next, and the next, and the next until you wake up in a world of fear and violence and wonder how we got there.

8) With the final systems in place, the Nazis started handing down the orders for mass murder to these groomed, promoted, pumped-up psychopaths and narcissistic sociopaths, who in turn forced the already brainwashed military into 'following orders'.

9) The result: by this stage anyone who spoke against the Nazi party was powerless. Those who opposed were arrested, charged with treason or some other like crime, and then imprisoned or killed -

10) It was too late to speak up or stop the Nazis because they had already built up a secret police and a paramilitary killing machine in the SS; and had re-written the law books to criminalize dissent. Dissent was no longer possible - the people had waited too long.

Recreating the Nazi state in the U.S. would be relatively simple - in fact, when we look at the comparisons, we a over half-way there - and completely believing in our cause - just like the German population under Hitler.

We are half-way there and we don't even know it.

Neither did the Germans.

Fear of terrorism.

False-flag attacks.
Mass patriotism.
Pre-emptive invasions based on those falsehoods.
A growing (unstoppable?) paramilitary killing machine
(the CIA)
Perpetual war.
Deception to cover the corporate objectives.
Re-writing of laws to criminalize dissent.

There is no 'conspiracy theory' here - you can see the pattern for yourself. The question is: what will happen when it turns on us? And in a generation, will we even notice?

"[F]irst they came for the Socialists, and I did not speak out - because I was not a Socialist.

Then they came for the trade unionists, and I did not speak out - because I was not a trade unionist.

Then they came for the Jews, and I did not speak out - because I was not a Jew.

Then they came for me - and there was no one left to speak for me." ~ *Martin Niemoller* [6]

Martin Niemoller endured over 7 years of torture in a Nazi

---

6. Holocaust Encyclopedia: Martin Niemöller (1892–1984) was a prominent Protestant pastor who emerged as an outspoken public foe of Adolf Hitler and spent the last seven years of Nazi rule in concentration camps.

concentration camp for *dissent*; he was too late - the laws against free speech were already gone, the paramilitary SS already in total control.

Are we heading into a future where protesting your
working conditions or joining a trade union or opposing a
draconian law change could see a drone hovering over your
family home while you are assaulted and arrested in front of your children?

Even though we can now tick off every key factor leading to Nazism in modern America, we still have no way of knowing how far fascism will go in the future Western world - the one thing that is certain though - is that drones *are* coming.

Meet the Fockers

The visible sentinels of the 'secret' Af-Pak war are the MQ-1 Predator and its replacement, the MQ-9 Reaper. Various stealth drones have also been displayed at air shows; and one was brought down and its wreckage recovered in Iran, apparently by means of cyber attack.

One thing to consider seriously though, is that drones will evolve; they will not stay the same.

Any piece of airborne hardware can be converted into a Remotely Piloted Aircraft - an Apache gunship for instance.

For domestic control, drones will need to hover over their targets, and will most likely be entirely new machines based on the multi-rotor-copters used by hobbyists and the paparazzi.

With hovering drones will come instant electronic identification. Now this may sound good to some; many I know already say: "I'm no criminal, so I've nothing to fear."

But the problem is, we don't know what will be considered 'criminal' behavior in the future - In Nazi Germany, it was standing up to your boss, or joining a trade union - which quickly morphed into anything and everything, like getting sick or opposing the government.

When a perpetual police-state comes, it will be final - you will never be able to change anything again because change and protest will be criminalized.

By definition you will be a slave.

Drones in the Middle East 'fighting' Radical Islam, is merely the first rung on the ladder to enslavement; but how far up that ladder will we go? And how far is too far?

Research author Jim Marrs puts it like this:

> [O]f particular concern to Americans who look beyond the advancing fascist agenda is the question of who will come to their rescue? Nazi-occupied Europe, and even many Germans toward the end of the war, looked hopefully to the Allied nations for their liberation. If America, today the world's foremost empire - a new Reich - falls under fascist domination, where can Americans look for deliverance?[7]

So without further adjure, let's meet the family.

---

7. Marrs. Rise of the Fourth Reich

MQ-1 Predator

MQ-9 Reaper

RQ-180 Sentinel - 2014

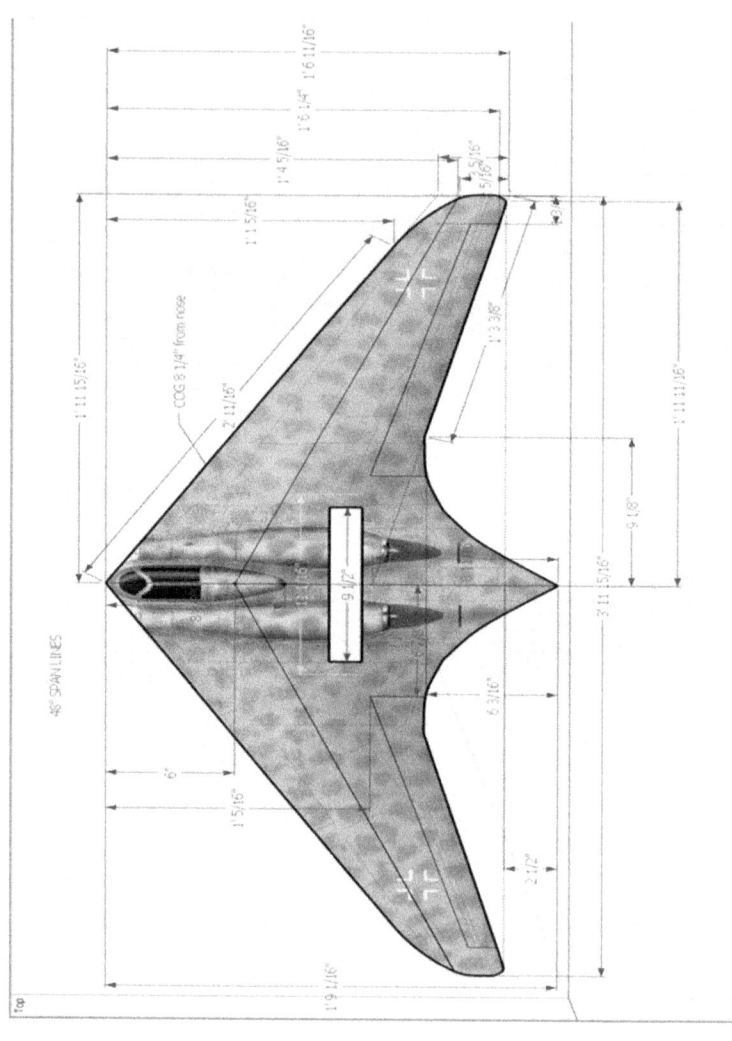

Nazi Horten Ho-229 - 1945

Although not a drone, the Horten-229 is included in the 'family'. Found by Allied personal at WWII's end, the 229 is clearly the forerunner to our modern stealth-wing design, bearing obvious resemblance to the B-2 Stealth Bomber, and the smaller RQ-180 Sentinel drone. The 229, along with a plethora of still-classified *Paperclip, Alsos,* and *Lusty* finds, shows that the Nazis were not only advanced in their thinking, but were on the cusp of breaking into an entirely different realm of technology altogether, as far back as the 1940s.

Concerning the Predator and Reaper drone systems, we already know much of their capabilities and mission parameters, but the newer RQ-180 is clearly different. The Sentinel is obviously a stealth drone that can fulfill ISR missions requiring both long loiter time, and non-detection.

If the RQ-180 carries a weapon, we don't know about it.

As stealth capability is not a requirement for aircraft survival in the Mideast, one wonders: why the need for such a specific stealth drone?

Several possible reasons exist:

1) The CIA (the drone's operator) does not want its *allies* knowing what it is 'surveilling'.

2) The CIA doesn't want the USAF to know where its drones are. ie: the CIA is conducting very

different missions to those of the Air-force, and most likely above USAF security clearance.

3) The CIA doesn't want Russian, Chinese, Iranian, or Asian radar platforms discovering its extracurricular operations.

4) It is 'weaponized'.

5) It is illegally crossing boarders, conducting ISR (and or strikes) over Iran, and possibly other countries.

6) It can assist in the accuracy and/or use of stateside scalar weapons platforms such as HAARP arrays.

Going West: Middle-Aged Spread

Suddenly and once again - out-of-the-blue - Islamic fundamentalism is spreading.

Drone bases have appeared in Africa to 'monitor' the spread. A drone base in Djibouti, a small country in The Horn of Africa bordering the Red Sea, gives the CIA drone access to Yemen, Somalia, and anywhere else in a drone's giant 20 hour operating circle. Of course al-Qaeda has spread to these countries.

But now Islamic fundamentalism has spread to West African countries like Nigeria and Chad, and likewise, this has elicited the spread of drone bases.

A very secret base has now appeared in Niger, West Africa. According to The Washington Post, this drone base is surrounded by a wall of razor-wire and security systems.

Drones from this new base, according to John Brennan (the first director of the National Counterterrorism Center, which shapes the overall strategy against al Qaeda and its allies across the 16 agencies of the U.S. intelligence community) are un-armed surveillance drones, monitoring

Islamic Fundamentalist spread in both West, and North Africa.

The paradoxes and problems with this 'spread' are as follows:

1) As we know and can see via both CIA behavior and outcomes from 1978 - 2015, IS and al-Qaeda type groups are kindled and controlled. If they migrate into Africa and other countries, it is because they are 'allowed' to.

2) The Islamic Fundamentalist mindset was seized upon by the CIA in 1978/79, and fed to the Mujahideen so as they would fight and beat Soviet armies, lured to Afghanistan by the CIA.

3) This mindset may have died out after the Soviet defeat in 1989, but instead, by exacting U.S./ CIA actions in Afghanistan and the Mideast, the CIA fueled Jihadist mindset morphed into 'Islamic Fundamentalism', as predicted by the Rosin Affidavit.

4) This eventuality justifies continued U.S. presence in the region.

5) Drones simply monitor the 'spread' of Militant groups in West Africa; drones do nothing to eliminate it.

6) To eliminate Rogue Terrorist groups, a very different approach is/was needed.

7) The spread of the IS mindset to other countries like the U.K. and Australia is impossible to stop via social media. People join causes, even crazy, heavily manipulated ones.

The big question is: is all this migrating and spreading of Islamic Fundamentalism being allowed for a reason?

Is the CIA's goal to inflame an international war?

Is a large, low tech, conventional WWIII a possible or desired outcome?

Or is this 'spread' simply allowed to further justify CIA control, and prolonged Mideast operations and occupations?

This is difficult to predict, but bear in mind the following.

1) With the televised atrocities committed by IS against Westerners, patriotism is being super-charged.

2) Via heavy media coverage, we have been made to deeply fear IS.

3) If IS is 'allowed' to spread by the CIA, (controller of the regional assets), then other countries' soldiers will be lured into a bloody war that could have been snuffed out months (or years) ago by the CIA alone - but instead has been inflamed.

4) A larger or global war would be hugely profitable for the U.S. military/ industrial companies that make the bullets, guns, bombs, and machinery.

5) The profit from such a war will be paid for in the blood of the men, women, and boys that we send into a clearly pre-planned war for profit.

6) The writing is on the wall. CIA behavior and outcomes in the Mideast and in controlling IS are heading to only one end - conventional war, but;

7) If conventional war for profit is a goal, first however, the Islamic Fundamentalist movement must be further grown: their numbers and strength are still too low to be a 'real enemy' in a conventional ground war.

This poses a very real and very frightening series of questions:

1) Can the CIA control the monster it has been creating since 1978?

2) Will the manufactured IS mindset become a global cancer without cure?

3) How do you stop a terrorist mindset, initially meant for Arabs, from spreading out of control throughout other countries?

If this happens, the CIA, created, funded, and allowed to exist by the U.S. administration, will be entirely to blame for fanning the Jihadist flames and using garden hoses to control the blaze.

Immediate solutions (none of which can possibly happen under our current system of government):

1) Remove all support for IS.

2) Investigate and cut off Saudi and other funding to IS and al-Qaeda. This should have happened in 9/11's aftermath, but was allowed to continue in ominous silence.

3) Remove the CIA's strike arm: they have become a paramilitary headless horse, with little oversight, total deniability, and are heading total control over our future.

4) Install a non-corrupted intelligence *gathering* agency; this is clearly a fantasy.

5) Handover *all* strike operations to the USAF, an organization with more accountability and oversight.

5) Wipe-out IS in a fast Blitzkrieg operation: the cancer is there now - it must be cutout before it spreads out of control. With no IS in Iraq and Syria, *no* new recruits will flow in to the region - literally cutting off IS' head.

6) One must ask why this has not been done - and one must see what is really afoot.

7) With the major political parties controlled and run by the colossal Military Industrial Complex, these changes will *never* happen under Republican, Democratic, or Conservative governments - this is why we never see any change of tactics, regardless of who holds office.

8) If we ever did see a radical change of government, the new officials would be threatened into compliance with the corporate MIC program. (You may use your imagination here). John F. Kennedy initially did the biding of the mighty MIC lobbyists, but the moment he went rogue and tried to undo their dirty work in regards to Cold War deception, pre-emptive invasions, and dismantling the CIA - he was removed.

8) So it is with sadness that the only real future, arrived at by behavioral deduction and the un-stitching of the Cold War deception montage, is that our children will go to the Mideast to fight and die in a bloody war using basic technology (to keep the weaker IS in the game) - a war that has been designed and created by the very governments that you elect into office - they will die as pawns in a con for corporate profit.

9) A spin-off of such a war will be continued domination of the Mideast oil ring - a perceived shortage of oil, and an up-stepping of the barrel price.

If we now go back a step, and consider basic Middle East deception, drones appear strong in the Mideast when they are actually weak - drones are vulnerable to air-defenses, signal jamming, and cyber-attack - drones would not survive in 'real war' conditions against a complete enemy; they are an ISR and policing tool. So to return to topic, we now must examine the *real* strength *behind* the drones; and ask even harder questions.

The Biggest Billy Goat: The Real Power Behind the Drones

Contrary to popular belief, it is not just the Predator and Reaper drones, circling the Mideast like Skynet's Terminators or the Sentinels of the Matrix' machine world that secures U.S. air-superiority.

Undisclosed and super-secret *un-conventional* weapons systems underpin U.S. overt and covert wars of conquest and/ or occupation.

Does U.S. power lie in mega weapons like HAARP arrays, which according to a Bernard Eastlund Patent, can deliver atomic sized explosions to any point on the planet or upper atmosphere?[8]

Another important question in regards to global superiority is: what is the real level of secret U.S. space operations? If an above top-secret 'black-budget' space program exists in the form of ultra fast anti-gravity discoid craft - then this massive technology gap would provide the power to dominate any opposition to U.S.

---

8. United States Patent №: *4,873,928*. Issued to Bernard Eastlund (at the time a consultant to ARCO industries)

imperial wars. But does this technology exist? And if so, is it ours? This however, is a question for later.

It is clear that Reaper hunter/ killer drones are a terrifying reality for the some twenty million inhabiting Pakistan's North-Western tribal borderlands. They will see drones overhead, or hear them coming, and not know whether a Hellfire missile will destroy their home or kill or maim their family - and this fear exists for both terrorists and innocents as CIA drones often hit the wrong targets. Peaceful civilians outnumber terrorists in this region in the order of tens-of-millions.

CIA drones often hit innocent targets, or as the State Department puts it, incur 'collateral damage', because drone operators use computerized 'behavioral targeting' software and techniques to ID 'terrorists' among the innocent.[2] In Af-Pak region of drone warfare, if you look a bit like a terrorist via the computer uplink, you can and will be exterminated by drone - innocent or not - and this is deemed okay because according to our leadership, the end justifies the means in reference to controlling terrorism. In the near future, it is envisaged that this process of computerized drone assassination could become entirely autonomous, or human-free - computers deciding who lives and who dies in the Middle Eastern Empire.

---

9. Turse & Engelhardt. Terminator Planet

Predator and Reaper drones are the visual or overt workhorse of the 'secret' Af-Pak war, but in a conventional military sense, are very weak.

1) Drones have no air-to-air, or self-defense capability - because they don't need it. There are no enemy aircraft or Surface to Air missile systems in the Mideast that could shoot one down.

2) Drones carry no pre warning systems or SAM detection sensors that would normally be onboard piloted CAS aircraft - because there is no risk to drones in the Mideast.

3) They are vulnerable to cyber attack. Data links flight control or visual systems could be severed by electromagnetic jamming. Wireless data signals could be hacked. Iran however, appears to be the only country prepared to do this.

But Reapers are a slow flying, fear inducing, means of corralling and controlling ever rising Mideast terrorist groups operating within Mideast boarders.

So the real reasons behind CIA drone operations become clear:

1) Drones fly the boarder areas of Af-Pak, and Iran/ Iraq using Hellfire missiles and laser guided bombs to contain the Islamic State to Iraq and Syria - the countries that need to be 'regressed' or kept in technology debt.

2) Drones strike civilians and women and children, to 'stir up' anti western mindsets, creating the terrorists of the future and pushing men to join groups like IS.

3) Drone pilots are unaware of this motive - CIA controllers operating above Top-Secret-Crypto simply employ defective targeting tactics and rules of engagement (like behavioral recognition) to ensure civilian casualties, and maintain plausible deniability - if you can even call it that.

The message is: The CIA is the operational arm of the Corporate Military Industrial Complex, which sits atop the pyramid of power.

The MIC needs continued presence in the Mideast so as to control its oil empire.

The CIA therefore needs terrorism.

If one considers Dr. Carol Rosin's affidavit, in which she stated that in 1974 Werner Von Braun told her that, "To create a climate of fear, first the Soviets will be considered the 'enemy'. Then we will use terrorists and third-world crazies", - then the CIA have followed Von Braun's script to the letter.

The CIA funds terrorist groups; they allow the Saudis to fund Al Qaeda and do nothing.

In terms of drones - drone presence and 'inaccurate' drone strikes turn peaceful citizens into 'terrorists', or at least anti-American enough to justify further 'policing'.

The CIA needs terrorists to achieve the MIC's corporate goals, and the MQ-9 Reaper is the tool of choice for corralling, controlling, and killing to keep IS militants on task.

Reaper pilots operate out of Multiple U.S. Air Force Bases including Creech AFB Nevada and Holloman AFB New Mexico. The CIA flies Mideast drones from its HQ at Langley. Takeoff and landing crews often operate from Khandaha in Afghanistan.

If the CIA wanted to wipe out IS in Iraq and Syria, they would not use drones; drones can't pack the required firepower.

It is interesting to note that every 'ex Air Force General' interviewed by the media rattles out the same line: "airpower is ineffective in Iraq because the militants hide among civilians."

At first this seems a reasonable assumption.

But this common statement lies at the heart of CIA Mideast deception for several reasons:

1) U.S. forces have disbanded Iraq's Air Force, making the USAF the only Airpower in the region. (The IDF is a U.S. ally.)

2) If IS moves to take a position, town, or any other objective, they will need both a means of organization, and to form a fighting force. As they

form, and when they move, airpower delivered by piloted fighter-bombers will decimate them, their vehicles, and weapons.

3) With low numbers, (around 5000 at full strength) when the IS remnants melt back into civilian populations or towns, ground forces can then sweep the town clean. This phase of course is the problematic one, but it is clearly necessary. Not all IS militants will be found, but with most of their force lost, what can mere individuals do?

4) With a decade of occupation in Iraq, the U.S. must have built a substantial intelligence network. Why do they not gather intelligence on IS' base locations and meeting points and then eliminate them?

Another tired line is: "Airpower has never broken the sprit of a country or won wars". This statement is also common, but really deserves a lengthy study, because it depends on the type of airpower, the tactics, and the ability of the aggressor.

Since WWII the USAF has ensured total, global air superiority - they have ensured this for obvious reasons.

The first thing that must be attained before any battle or invasion can be fought - is air-superiority.

During WWII the combined British/ U.S. saturation bombing of German cities did not cause

German capitulation, but the Allied motive was 'punishment'.

Although not aircraft, Hitler's V1 and V2 'vengeance' rocket attacks on London killed many civilians, but was never big enough to affect the War's outcome - it was as the name suggests - merely Vengeance.

Consider this: in 1945, two aircraft delivered two bombs to two Japanese cities - ending the Pacific war, nullifying the need for a bloody D-Day style invasion of mainland Japan.

Today air superiority allows the U.S. to strike freely and at will.

Drones do not provide air-superiority, they merely operate under its cover.

The pre-eminent ways to achieve air superiority are to:

1) SEAD: Destroy or render enemy air defenses in-operative, such as radar and (SAM) Surface-to-Air-Missile sites, and;

2) Mass 'vulching': To reduce, then eliminate an enemy's air interception ability by destroying 'his' aircraft whilst still on the ground.

In the modern age, or post Cold War age of secret *secret* wars, what kind of weapons platform provides this ability? And who does it protect against? Are Russia and

China the real enemy behind the enemy, or are they secretly on our 'payroll'?

One asks: does Nuclear First Strike provide the power to dominate? - No, because of early detection and possible nuclear retaliation.

Is it ICBM Strikes? - Maybe, but ICBMs are clearly vulnerable to missile defense systems and early detection.

Cruise Missile Strikes are more strategic and don't deliver the punch required for total domination, and are once again vulnerable to air defense systems.

Strategic Bomber Strikes are also far too vulnerable to Enemy Air Defenses. Stealth Bomber Strikes don't provide the required punch.

Is there even a single weapon that gives the U.S. the power and confidence to claim global dominance?

One very important point regarding the Middle East and CIA deception operations in the region, and something that we have not yet explored is this: the media is both fed and drawn to Mideast terror footage. This singular focus (like the singular focus that the disappearance of MH-370 provided for several news days) provides a diversionary effect.

While the world has IS rammed down their throats, is something else happening elsewhere that needs to remain hidden from our view?

Is ISIS just a diversion?

And if so, a diversion to what exactly?

To appreciate the extent of the world's underlying secret weapons programs, and to highlight the pointlessness of Mideast drone strikes at all, we need to explore the rise of a very different machine: the scalar weapon.

Up-Scaling:
Scalar Weapons

"Others are engaging even in an eco type of terrorism, whereby they can alter the climate, set-off earthquakes, and volcanoes, remotely through the use of electromagnetic waves . . . It's real - and that's the reason why we have to intensify our efforts." [10]

~ William Cohen. Defense Secretary, 1997

Talk of scalar weapons raises both eyebrows and questions. The development and use of electromagnetic (EM) or 'beam' type weapons which act over great distances with varying degrees of destructive power, are rarely admitted to or discussed.

Do they even exist at all?

Clearly, as of at least 1997, Defense Secretary Cohen, and therefore both the Pentagon and the CIA, are aware of such possibilities - at least in theory.

Some claim that Cohen's statement has been taken out of context, but Cohen clearly stated *"It's real"*, and when coupled with Zbigniew Brzezinski's statements of a desire to achieve 'weather modification weapons', and the many observed Soviet EM tests, Cohen's statement shows both belief and knowledge of EM weaponization. We still tend to think of the things Cohen and Brzezinski mention as the stuff of

---

10. http://www.defense.gov/transcripts/transcript.aspx?transcriptid=674

'conspiracy theory' - but those days are in reality, long gone.

If we go back to the start however, the first major points to consider in answering the questions surrounding EM weaponization are as follows:

1) There exists considerable proof of Nazi scalar weapons development during WWII.

2) There exists considerable proof of Soviet scalar weapons development, initially under Stalin, but also later under Gorbachev and Putin.

3) Before 1945 and Nazi capitulation, elements within both the Soviet Union and the U.S.A prioritized the capture of Nazi technology, their scientists and equipment at WWII's death - and the 'cover' of these operations and acquired technologies by a clever and vast montage of deception.

4) This deception either inflamed inter-superpower paranoia, triggering a secret Cold War within the 'nuclear' Cold War, or;

5) Was the launch of a joint Soviet/ U.S. deception that would pit the two as 'enemies', whilst they worked as *one* to enrich their military Industrial Complexes, suppress technology to maintain and continue Oil Company profit domination, and develop advanced propulsion and electrical generation technology without public knowledge?

The fact that scalar weapons were envisioned during the forties, captured and kept secret by the Cold War 'super-powers', and then 'not talked about' and 'denied' whilst giant phased array antennae farms sprung-up around the globe, may be an indicator of a high-level technology cover-up.

Questions arise:

1) Do the giant billion watt HAARP arrays provide the overseeing 'muscle' which allows the CIA to operate its weak Mideast drones with little or no international response?

2) If the Russian Federation (and more to-the-point, the former Soviet Union) developed and possesses scalar electromagnetic weapons, why didn't they - during the height of the Cold War - initiate a first strike against their 'arch-enemy' - the West?

3) Were the USSR and the U.S. really enemies at all?

In January of 1960, in a presentation to the Soviet Presidium, Nikita Khrushchev (Premier and First Secretary of the Communist Party) said:

> ["We] have a new weapon, just within the portfolio of our scientists, so to speak, which is so powerful, that if unrestrainedly used, could wipe-out all life on Earth."[11]

---

11. Bearden. Fer De Lance

Questions immediately arise:

1) One instantly asks: why was this weapon not used against the CIA and Mujahideen in Afghanistan between 1979 and 1989, to stop the 'Soviet Collapse'?

2) Did Khrushchev's weapon fall through? Over the decades, strange, massive weapons tests deep inside the Soviet Union, observed by Vela satellites, and multiple international airline pilots, which fit the descriptions of no other known weapons platform, suggests not. (A point which we will explore later.)

3) Was the piping of Russian oil to U.S. oil companies at the time deemed to be more financially important by both Soviet and U.S. governments? Many countries would not buy Russian oil during the Cold War years, and many believe that said oil was being piped, shipped, and sold for greater profit as West Texas Crude, whilst America was 'supporting' Afghanistan in a ten year war against their Soviet 'enemy'. (Again, a point to be explored later.)

4) This raises the question: was the *real* purpose of the Soviet/ CIA action in Afghanistan from 1978 - 1990, not to destroy the Soviet Union at all, but to 'create' Middle Eastern terrorist groups as described by Dr. Carol Rosin in her infamous Rosin Affidavit? A Military-Industrial civilization always needs an enemy, or the threat of an enemy, or it (the MIC) will financially

collapse. If military necessity is the backbone of a nation's finance and industry, perpetual war is integral to the nation's immediate financial survival.

If no real wars or enemies exist, they must be, and will be, fabricated. The psychology of deception then twists us to believe a false reality. The false reality then, over time as it has, will become a *real* reality, but one that is weak and must be made to appear much stronger and more dangerous by media driven availability cascade.

Communists, Terrorists, Asteroids, and Aliens are the enemies that Werner Von Braun, (Paperclip Nazi and Director of the Marshall Space Flight Center), told Dr. Rosin would be created and used, one-by-one, as the 'enemy' that would keep the funds flowing into MIC coffers, and to justify 'space based' weapons.[12]

But were the Soviets 'in on it' all along?

Stalin was portrayed as totally untrustworthy by Churchill and Roosevelt, but were 'The Big Three' part of the same organization all along?

Was the entire Cold War a staged money siphon for a borderless, international secret organization?

If not, why did the former Soviet Union test, but never use, electromagnetic weapons against the 'West' in Vietnam and Afghanistan?

---

12. The Affidavit of Dr. C. Rosin. Taken from the Disclosure Project website

Why did the USSR allow itself to capitulate?[13]

What kind of secret-secret war is really going on?

To even begin to answer those questions, we must first take in a brief history of Cold War scalar weapons development.

In discovering the true nature of Scalar weapons, two authors, each brilliant in their own right, standout as the leaders in the field: Nuclear Engineer and physicist, Lt Colonel (Ret) Tom Bearden, and Dr. Joseph P. Farrell. Farrell focuses, as he does, on the Nazi links, whilst Bearden focuses on covert Russian development and continuing Cold War type threats to Western civilization.

But are they both missing the key point - Soviet/U.S. allegiance, and mass Cold War deception?

Before we analyze that bizarre question, let's examine some of the history.

---

13. The Soviet Military capitulated in 1989; however the dissolution of the Soviet Union was formalized in 1991

A Brief History of Secret Physics

Zero Point Energy (ZPE), Quantum Flux, Quantum Energy, Quantum Potential, The Unified Field, The *Ether*.

These are names used at various points in human history to describe the source of energy thought to underlie all life.

Other cultures have words like chi. The Chinese use qi.

Pre World War Two, in the dawning era of Quantum Theory, the world's cutting-edge scientists worked under the idea that the 'Aether', a limitless source of energy simmering below the Plank Scale, was the force unifying everything.

James Clerk Maxwell used the Aether to account for the electromagnetic field produced (seemingly from nothing) by an electric charge. For without this theoretical ether, electromagnetic fields break the 1st law of thermodynamics - that 'something' cannot come from 'nothing'.

When one studies the subatomic 'dance' of protons, neutrons, quarks, and hadrons etc - these subatomic particles constantly appear from 'nowhere', carrying minute electrical charges - they also change from one to the other at will.

If, (like post WWII physics says), there is no ether, what is the source of these curious subatomic charges, and their electromagnetic fields?

Necessity - The Mother of Weaponization

World War Two's escalation saw the 'secretization' and 'compartmentalization' of weaponizable physics for military purposes, in both Nazi Germany, and the Allied countries - and at the exact time in history when Quantum Theory was emerging as a powerful force in both theoretical and *'engineerable'* physics.

Behind the veil, WWII was a quantum war. After WWII the veil became a steel curtain of secrecy, held-up by an increasingly complex montage of deception which had to keep pace with the rapidly advancing (but entirely secret) technology.

After WWII's final shots, Einstein's earlier theories of Special and General Relativity were pushed to become the accepted 'mainstream' theories of physics. Relativity abandoned WWII's Ether theory. In fact, Relativity disproved the Ether's existence whilst becoming 'fuzzy' and paradoxically triggering the need to 'unify' both the celestial and the sub-atomic worlds.

How did such a pervasive idea - the idea of the Ether - suddenly become extinct?

Is the Ether real or not?

Was Ether theory torpedoed, obfuscated, or covered by the deception montage, so as to protect secret technology?

Is there, or is there not, a 'life-force' driving the mysteries of the sub-atomic world?

Has Ether theory been discarded by educative scientific institutions only to be continued in secret by a small section of the post WWII Military Industrial Complex?

Is there a sinister side to quantum physics?

These questions engulf Quantum Theory, and as we shall see, the rise of secret weapons as well. Quantum Theory has given us much, but what it shows us is that we do not yet know reality.

We don't know how the world works.

As yet the human species does not know exactly what life is, how it started, or where it came from. We do not know whether we are an illusion, a dream, a shadow of another reality, or whether we live in a multi-dimensional universe.

Are we not smart enough to work it out?

To be sure, we have had no choice but to walk a long evolutionary process in order to discover and understand the little that we do know about the world in which we find ourselves.

Will we know the answers to these questions in our lifetime?

Some may. In WWII's immediate aftermath the necessity of both self defense and global domination drove the scientists of the Soviet Union and the U.S.A. to focus their intense efforts on weaponization. From a Soviet standpoint:

1) The U.S. had gone thermonuclear. (Stalin would not do so until 1949).

2) Russia's cities and people were smashed and broken. Over 25 million Russians had been killed during WWII.

3) A war footing had to be maintained which drained all Russia's resources.

4) A weaponizable physics had to be found to put the Motherland back on an even footing with the U.S.

5) Like the U.S. Soviet Intelligence had captured Nazi scientists. Stalin rapidly adopted and continued the Nazi approach of considering and probing every conceivable physics to achieve military dominance.

Now that opening foray comes across as a little mystic, but before we tackle our bigger questions, we firstly need to peruse the basics of Quantum Theory?

Quantum Theory 101

Quantum Mechanics is the study of the sub-atomic world.

The sub-atomic world is studied to answer fundamental questions like: how did life start? What is consciousness? How is the universe made up? What is our reality? It was also studied by the Nazis as a means to terrific weapons of mass destruction.

In the early 1920s the world's brightest minds began formulating physical experiments that would show the behavior of the subatomic particles making up an atom (such as electrons, protons, and neutrons). Bizarre laws that defied belief were derived from these experiments.

The Copenhagen Interpretation blew their minds. In 1927, led by one Neils Bohr, a group of emerging Quantum Physicists saw that an electron was not just a solid particle, but also took the form of a 'wave', like light.

Observer Effect was born, where an electron will be either a physical subatomic particle, a wave, or both at the same time depending on how it is observed.

Being in all possible locations or outcomes at once, the electron is said to be in *super-position*. The act of recording or observing will collapse the 'wave' and cause the electron to again become a physical subatomic particle.

They learned that mere observation affects the outcome of experiments - or reality. The contemporary evolution of this idea is that 'belief' affects the physical world.

Many incredible contemporary theories have come out of Quantum Physics such as super-string theory, and multiple-universe theories - but the further down the quantum rabbit-hole they went, the more bizarre things got. Quantum physics failed to unify gravity and electromagnetism.

But over *two hundred* years before Einstein released his theory of Special Relativity, a man called Boscovich had already published an idea light-years ahead of its time - Boscovich had published a theory that unified gravity and electromagnetism - a Unified *Field* Theory.

The Suppressed Theory

In 1943, when Nazi physicists were turning every stone in their hunt for a weaponizable physics, the cutting edge theory *was* Boscovich's Unified Field Theory, re-packaged by Albert Einstein between 1925 and 1930.

Nicola Tesla, who is rumored to have designed the electrical equipment for the Philadelphia Experiment, is said to have worked under this unified field theory which unified gravity and electromagnetism mathematically, and in a way that could be used practically.

If Tesla and Einstein were involved in the Philadelphia Experiment (and evidence suggests they were both in Philadelphia and employed by the US Navy at the time) it is odd that this strange theory underpinned the experiment's physics; a theory that disappeared after the War, never to be heard of again in mainstream physics. Now it could be that Einstein's UF theory was discarded legitimately, but as we shall see, there is a strange and sinister history surrounding the theory - a history that draws a researcher to investigate further.

Is this 1928-30 Unified Field Theory of Albert Einstein's, borrowed from Boscovich, a secret that has been hidden and protected for decade upon decade in order to conceal the Philadelphia Experiment, and possibly later experiments in torsion, space/time manipulation, and occult ether physics?

If so what else does such suppression conceal?

Technology suppression must occur. Suppression of technology keeps the U.S military ahead of its enemies. Deception must surround such suppression and we, the public, must accept a certain level of such suppression and deception - mustn't we?

But as always, where there is technology suppression there is a suppressor - and where there is a suppressor, there is a secret.

Making Waves: Scalar Waves

According to Colonel Tom Bearden, scalar waves are waves in the electromagnetic spectrum, which however, fall outside of our 'current' theories of physics.

In short, Maxwell's equations used a mathematical language (quaternion) which was later discarded in favor of Heaviside's 'vector analysis'. Vector analysis, (and more importantly the conversion of Maxwell's equations into vector sums), does not explain or account for the 'source charge' carried by subatomic particles, where as Maxwell's discarded equations did so.

Under currently taught physics, the charge held by electrons, protons, neutrons etc, has no source - the charge just is. Maxwell's source of energy however, lay in what was then called the 'vacuum'. Many suggest that the idea of this Zero Point Energy was covered up by the superpowers in the 1930s when they realized virtually anyone (including enemies) would inevitably gain the power of mass destruction via such physics.

Nikola Tesla used such physics in his advanced electrical engineering, denouncing 'Einsteinian' theories of gravity and Relativity. Tesla of course, was the genius behind our current electrical age[14] (which has not changed since its invention) - the system of electrical transmission we have now was not supported by the electricity monopoly of the time because less efficient systems would have been more profitable - the most efficient system - Tesla's goal of wireless electricity to every home and building, would've been devastating for the monopoly.

Tesla believed in the Aether, now referred to as Zero Point Energy (ZPE). Before the First World War, Tesla had envisioned a 'beam weapon' which back then he called a 'death-ray'. He tried to sell his 'death-ray' to the U.S. Military, who at the time, could not believe or grasp the concept.[15]

Bearden believes that longitudinal electromagnetic waves, explained by Maxwell's discarded mathematics and Whittaker's electromagnetic theory of 1903-04, along with Tesla's evolving ideas, were discovered by the Soviet Union after WWII, and were hence engineered to create the first scalar weapons.[16]

---

14. Lyne. Pentagon Aliens
15. Ibid
16. Bearden. *Historical Background of Scalar EM Weapons (1990)*

The Nazi Connection

Bearden and Farrell agree that during extensive WWII research into Over-The-Horizon (OTH) radar development, German scientists 'stumbled' onto scalar anomalies and were able to recognize a weaponizable potential.

Farrell in his *The SS Brotherhood of the Bell*, uncovers a German patent diagram housed in Germany's Munich Museum, showing how the *Freya* OTH radar array was to be used as a guidance system for inter-continental rockets.[17]

Freya's three antennae sent 'out-of-phase' pulses, (or waves in which the respective signals feeding her antennas were varied in such a way that the array's effective radiation pattern was reinforced in one direction and suppressed in all other directions).

This 'phased array' type of antennae tuning 'bends' its microwave signal, thus allowing it to 'look' over-the-horizon and guide a missile to its target beyond line-of-sight parameters.

---

17. Farrell. The SS Brotherhood of the Bell, pg229

The obvious Nazi need for this technology was Hitler's so named Amerika Rocket - a much desired ICBM that could strike the Mainland U.S.A from Nazi controlled Europe - such a rocket was under late war development by Nazi rocket engineers.[18]

The point here is: because phased array systems underpin Tesla's scalar weapons theory, Farrell draws the obvious conclusion - that the *Freya* radar scientists (knowingly or unwittingly) detected non-linear EM effects in their research, and were themselves bent along a new path - a path leading over a new horizon to EM weaponization.

What is known as a Tesla Howitzer, (a-k-a, a Scalar interferometer) works in a similar way to the old Nazi OTH guidance-radar. The difference is Tesla's beam weapon fires out-of-phase pulses, sent at differing times and at differing velocities, but which arrive at their intended target simultaneously, causing destruction or some other intended result.[19]

> [It] would have been a small step for the Germans to make the change, and, as has already been indicated *this appears to be exactly what happened in their late-war radar experiments . . .* [20] (italics original)

---

18. Ibid. Farell infers that the development of such an ICBM suggests it was intended to be coupled to an atomic warhead, for what other reason would there be to go to such lenghts when clearly a conventional warhead would be pointless, as was the case with the V1 and V2 attacks on Greater London.
19. Ibid
20. Farrell. The SS Brotherhood of the Bell, pg231

Farrell cites several circumstantial confirmations alluding to the Nazi Germans as being the first to discover scalar weapons potential in phased array microwave pulsing; but a significant clue of Farrell's finding is this:

> [F]ourthly, there is the presence of one of the chief scientists involved in these radar and scalar weapons projects, Dr. Hellmann, both in the German, Soviet, and subsequent Brazilian programs.[21]

The point here is clear: when conducting their own version of *Operation Paperclip*, namely the locating and seizing of critical *secret* Nazi weapons technology, the Soviet Intelligence apparatus seized upon the Nazi *Freya* scientists and equipment, whisked them back to the Motherland, and one-way-or-another, learnt everything the Germans knew of scalar weapons potential.

─────────────────────

21. Farrell. The SS Brotherhood of the Bell, pg240

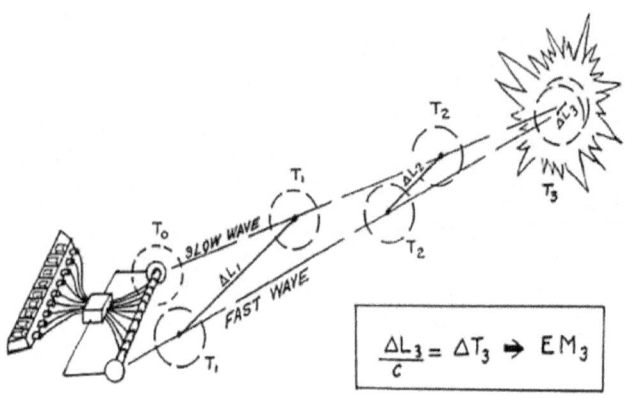

Tesla's Howitzer

A Cold Day in Hell

Lt. Colonel (retired) Tom E. Bearden, is a genuine physicist and held a high-level position as an advanced weapons systems engineer within the 'black' world of U.S. secret weapons development. His military areas of expertise included: air-defense systems; anti-radiation-missile countermeasures; technical intelligence, tactics and operations; and nuclear weapons systems and deployment, among other areas.[22]

Using his vast knowledge of history, weapons development, and deployment, Bearden has explicitly warned U.S. leaders of the threat posed by Soviet/Russian Electromagnetic Weapons.
In his book, *Fer de Lance,* Bearden uncovers the history of Scalar electromagnetic interferometry and its testing by the former Soviet Union. He also highlights the United States' obliviousness to such technology, and its un-preparedness towards scalar attacks.

---

22. chenier.org

We will examine Bearden's work soon, but first some burning questions and clear points of contradiction.

1) As of the 1997 statement by Defense Secretary Cohen, that EM 'weather weapons' are in use by 'enemies', and similar statements by Zbigniew Brzezinski, the assertion of U.S. ignorance is clearly over.

2) If the U.S. Intelligence apparatus was oblivious to EM First Strike back in 1960 - the heart of the Cold War, when fear of 'nuclear' annihilation was *meant* to be manic - why was no First Strike EM attack ever launched against the U.S. by the Soviets? EM weapons at this time, according to Bearden, could have rendered the U.S. nuclear 'dead-man' response (or its systems of Mutually Assured Destruction) inoperative by destroying the very circuits that controlled the systems. This suggests that the 'nuclear threat' or 'nuclear Armageddon' was false - designed solely to make the taxpaying population fear 'communists'.

3) Note that while no massive Soviet First Strike came, low level EM attacks against U.S. installations and citizens have been continuous since the 1960s. The U.S. Embassy in Moscow was bombarded by EM signals for decades, without staff being warned, whilst the CIA 'observed' and did

nothing. Also the Russian 'Woodpecker', an EM signal broadcast at the Mainland U.S.A. from within Soviet Russia, was known about but never actioned.[23]

These events drew Bearden to the conclusion that U.S. inaction amounted to proof of their inability to detect Scalar signals within the EM carrier signal, and was therefore proof of U.S. ignorance of scalar EM potential.[24]

4) I firmly believe (for reasons that will become apparent) that the U.S. has held implicit knowledge of EM weapons from the beginning (pre-WWI), and has therefore *never* been ignorant to EM attack - and has *allowed* 'low-level' EM 'test attacks' against its installations for either of two reasons:

a) To effect a 'Coventry' type deception, in order to fool the Soviets into thinking like Bearden - that the U.S. has no knowledge of EM weapons.[25]

This as we shall see is wrong, as U.S. sources have clearly known about EM weapons potential via Tesla since before WWI.

---

23. Beardon. Fer De Lance
24. Ibid
25. (A possible idiom). It is said that that Winston Churchill (with prior knowledge of the German bomber attack) let the English town of Coventry fall to German bombers, without warning them, to protect his intelligence gathering aparatus prior to D-Day. Some say this is false. Some say it makes sense. Some still say - we've put him in 'Coventry'.

Which opens the door to the unthinkable, but historically repeated phenomenon, of;

b) The U.S. powers have *allowed* Soviet EM tests against U.S. citizens to learn the prolonged effects on the unwitting human participants. This however, reinforces the growing idea of a Soviet/ U.S. alliance throughout the Cold War - showing up the entire Cold War, and all of its 'hot-wars', as falsehoods - a complex ruse to fool the world who paid for the whole thing via a colossal tax siphon.

5) The Russian Federation wants to see the end of the 'oil-age' about as much as the U.S or Saudi Arabia - they all make trillions from the industry.

In regards to the above, exactly what EM tests were carried out by Soviet scientists, and what exactly was observed? Such a question returns us to Bearden.

Project Vela

In Spanish, Vela translates to *He watches over.*
Project Vela was started in 1959 by the U.S. as the
first high-tech space-based snooping program. By
1963 the project had developed the sophisticated
Vela-Hotel and *Advanced Vela* satellites (six of each) to
specifically monitor the Soviet Union from 73,000
miles 'above' Earth.[26] Project statements say Vela's
main purpose was to ensure Soviet compliance to the
1963 Limited Nuclear Test Ban Treaty, by watching
for both space, and atmospheric nuclear
detonations.[27]

    The Limited Test Ban Treaty was signed by
The U.S.S.R, The U.K, and The U.S. The Treaty
prohibited all test detonations of nuclear devices with
the sole exception of underground test detonations.[28]
J.F. Kennedy said its purpose was to "slow the arms

---

26. http://heasarc.gsfc.nasa.gov/docs/heasarc/missions
27. U.S. State Department: Limited Test Ban Treaty pdf
28. Ibid

race"[29], but later information revealed that all were deeply concerned by the global spread of radioactive fallout, and a massive spike in carbon-14 levels in both hemispheres.[30]

Several obvious oddities exist regarding Vela:

1) Vela's strange apogee. Vela's orbital apogee of 73,000 miles put it well outside the Van Allen Radiation Belts (1000 - 60,000 miles above Earth), and almost $1/3$ the distance to the Moon (an average of 238,855 miles), an incredible distance.

2) The first satellite ever launched (Sputnik, USSR) was only launched in 1957. If Project Vela had been operational since 1959, it was an incredibly technical launch in many respects, especially when one considers the launch of Explorer 1 had only happened a year earlier.

3) Vela's distance from Earth, its secretization, and its early rush to deployment pre-1963, shows a clear need to 'watch' near Earth *space* at this time. This is stated in the 1963 literature, but must have been thought unusual at the time.

4) During the 1960s, nuclear detonations in the

---

29. 1963 Speech on The Limited Test Ban Treaty between the Soviet Union and the United States of America.
30. U.S. State Department: Limited Test Ban Treaty (pdf)

upper-atmosphere were carried out by both the U.S. and the USSR, causing the formation of several new radiation belts, un-controlled fallout, and who-knows what other horrific side-effects to the Earth's delicate balance. In saying that however, Soviet *nuclear* tests could have been detected in a number of other ways, which leads one to ask two questions:

> a) Did the U.S. Intelligence Community know and suspect Soviet EM weapons manipulation of the Earth's ionosphere at such an early stage? Soviet Ionosphereic Heaters are well known today - our own HAARP arrays are often called ionosphereic heaters, because they 'cook' or 'lift' parts of the upper-atmosphere in order to create 'mirrors' from which to beam or bounce signals.

> b) Was Project Vela watching for 'other' anomalies; such as human operated spacecraft?

These questions aside, Project Vela observed some very interesting anomalies.

The 1979 Vela Incident

Despite having previously detected a total of 41 confirmed nuclear detonations, an incident detected by Vela Satellite 6911 on September 22, 1979, raised a disturbing mystery.[31]

Vela 6911 was launched in 1969. Her photodiode sensors were designed to detect the two short, bright flashes (1 millisecond apart) preceding a nuclear fireball. On September 22, around 3am local time, Vela 6911 recorded two ultra bright flashes of light - seemingly consistent with another nuclear test.[32]

As (Soviet) above-ground nuclear testing was not allowed under the LTBT, an investigation was required.

U.S. Naval acoustic equipment (SOSUS hydrophones) detected an odd 'thump' at the same time and put the location near the tiny Bouvet Island, possibly the remotest island on Earth, about halfway

---

31. Lewiston Morning Tribune, 1979
32. Ibid

between South Africa and Antarctica. (Some information suggests Marion Is)

The Intelligence Community immediately pushed the idea of a 2-4 kiloton nuclear device as being the source. But no nations came forward.

Political maneuverings at the time had many speculating on a secret Israeli/ South African nuclear test, which seemed plausible (as it was known Sth Africa was working on a device) - but there were inconsistencies.

1) Vela 6911 recorded the two distinct flashes of 'light', but their velocities and intensities were inconsistent with a nuclear 'double-peak'.[33]

2) Vela 6911's EMP sensor was apparently inoperative, so 6911 was unable to confirm the distinct electromagnetic pulse emitted by nuclear detonation. This also means 6911 was unable to detect any other type of EMP signature.[34]

3) Oddly, a second Vela satellite recorded nothing.[35]

4) The second 'flash' was in the infrared spectrum, inconsistent with a nuclear blast.[36]

---

33. Bearden. Fer De Lance
34. Vela Data
35. Ibid
36. Bearden. Fer De Lance

5) Documents declassified after the fall of South Africa's 'Apartide' government show that a nuclear device did exist, but was not completed until November, 1979, two months after the incident.[37]

6) USAF research aircraft were immediately sent to the location armed with equipment to detect the radioactive dust and telltale products of nuclear weapons detonation; but none was found - the search area was radioactively normal.

7) At the time of the disturbance a strange and unusually strong auroral glow was observed in the upper-atmosphere from Syoaw Base in Antarctica.

8) The Arecibo Observatory in Puerto Rico recorded an unusual, fast-moving ionosphereic disturbance; not associated with ground based nuclear detonation.

9) It seems that the period between 1977-1980 was very different indeed, with over 600 loud 'booms' many accompanied by 'flashes' being recorded above the mainland USA during said period, triggering President Carter to ask for a 'full report'. These events were not nuclear explosions, and have mostly been put down to 'super-lightening'. These events were also detected and recorded by Project Vela.

---

37. CIA factbook

10) A similar 'double-flash' event was recorded by Project Vela in December of 1980. As no known natural phenomena can account for a 'non-nuclear' double flash, and as Vela's data shows the second flash being in the IR range and therefore inconsistent with nuclear detonation, a frightening series of questions are raised.

11) A scientific team of distinguished physicists investigated the 1979 Vela incident and (despite being coerced to deliver a 'nuclear conclusion' by the Intelligence Community) refused to conclude the event was the result of nuclear detonation. Their main reason was the Vela light flashes themselves not being those of a nuclear explosion. The second reason was the total lack of a radioactive signature.

12) Lightening and meteor strikes were ruled out as causes, as the Vela flashes were over 400x more powerful than the largest and brightest lightening flash ever recorded. (In a period of very bright flashes.)

13) A Stanford Research Institute group categorically ruled out meteoroid impact.

14) Evidence supporting nuclear detonation relied on CIA 'leaked' information, suggesting a South African test; which shows signs of the CIA leading us to the conclusion of their desire. Also, thyroids removed from Australian sheep containing trace

Iodine-131have been held up as evidence; but this has problems:

> a) No radiation or radioactive dust was found in the pin-pointed 'test' area, or in the vast distance between the test area and Australia.

> b) I^{131} can and does get into the thyroids of mammals from the mining of natural gas, and hydraulic fracture mining (fracking). Australia is of course, a heavily mined continent. Trace I^{131} could also have been a product of French nuclear testing in the South Pacific. With an 8-day half-life, I^{131} can remain in a sheep's thyroid for over a month.

It seems the CIA wanted us to think this event was a 'bomb', and that the other 600 odd similar events were a rare 3 year period of 'super-lightening'. What were they steering us away from?

15) If the expert scientific investigation team thought the 1979 Vela event was most likely a nuclear event, but just lacking evidence, they would have said so; but they did not. The light flashes were from something else. However, two 'unnatural', ultra bright flashes still occurred within Vela's detection range, along with an unnatural aurora, and a light-speed ionosphereic disturbance.

What was the 1979 Vela Event?

Nuclear physicist T.E. Bearden suggests that the signature, although similar to a nuclear explosion, was that of a scalar electromagnetic interferometer.

> In pulsed exothermic mode, [s]calar EM pulses meet at a distance, where their interference produces a sharp electromagnetic explosion, and hence a 'flash' similar to the initial EMP flash of a nuclear explosion . . . Prompt absorption and re-radiation of energy from this sudden plasma may be expected to present nearly the same 'double peak' profile as does a nuclear explosion . . . Note that the second flash detected was primarily in the infrared, almost certainly ruling out a conventional nuclear event.[38]

Could another type of 'secret' bomb have accounted for the event? It appears that the only weapon capable of producing all of the 'symptoms', is Bearden's scalar interferometer, whether Soviet or U.S.

But there's more . . .

38.Bearden. Paper: *Historical Background of Scalar EM Weapons (1990)*

Travis Stone

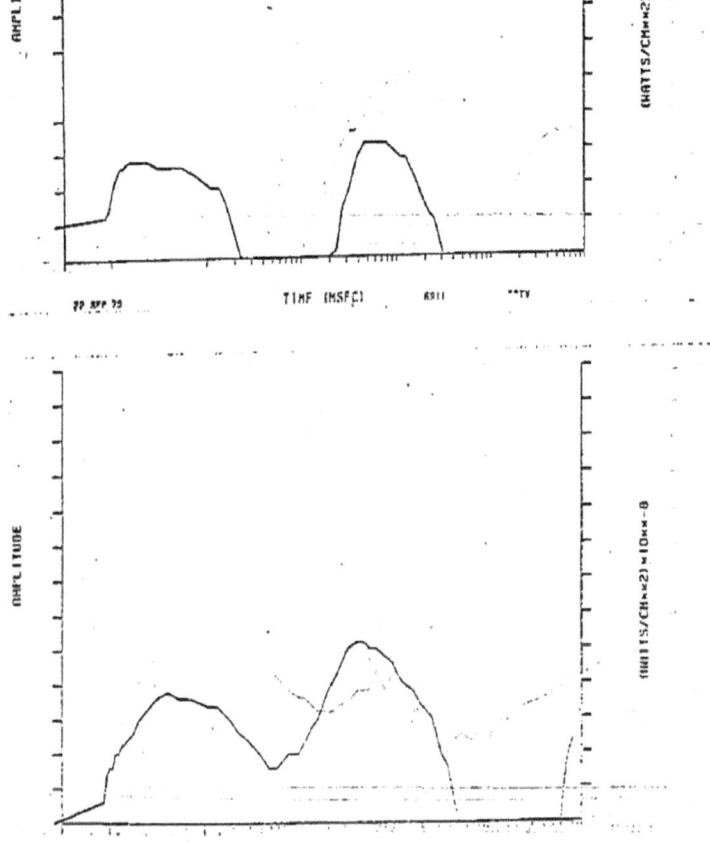

Vela Data, showing the 'double-peak'

The Soviet Woodpecker

In July of 1976, the year preceding the anomalous period of 'booming' and 'flashing' over North America, a strange, powerful signal suddenly started broadcasting in the SWR band, sending out a rapid tapping noise at 10Hz.[39]

The signal was identified as coming from the Soviet Union. Giant tower arrays at Chernobyl, Chernihiv, and in East Siberia were identified as the 'Woodpecker's' source, and were code-named *Steel-Yard* by NATO.[40]

The system was aimed directly at America.[41]

Many at the time (1976-1989), suspected sinister motivation behind the sudden appearance of Woodpecker; behavioral control, and weather manipulation were considered possible motives.[42]

The system was certainly not covert; anyone could hear its distinct 'tapping' in the communications

---

39. wikipedia. Russian Woodpecker
40. Ibid
41. Ibid
42. Ibid

band.

It continued however, unabated until the Soviet military's collapse in 1989.[43]

U.S. Intelligence established that the Woodpecker array was actually an OTH radar, although a giant and incredibly powerful one, operating between 10-40mw.[44]

As J.P. Farrell shows in his work *The SS Brotherhood of the Bell,* early Nazi OTH radar development was the forerunner to Tesla EM weapons theory. This point drew Tom Bearden to the conclusion that Woodpecker's SWR signal was possibly a carrier signal for difficult to detect EM scalar interference waves.[45]

Bearden believes that Woodpecker was indeed used to heat and cool areas of the upper-atmosphere over North America, (similar to modern ionosphereic heaters), in an attempt to control U.S. weather.

Interestingly, on May 18 of the following year, 31 countries, including the U.S. and the U.S.S.R. signed a treaty which forbade attack by, or the causing of, man-made storms, earthquakes, tidal-waves, and drought.[46]

---

43. Ibid
44. Bearden. Paper: *Historical Background of Scalar EM Weapons (1990)*
45. Ibid
46. Ibid

Today we tend to view weather-war as 'conspiracy theory', but Zbigniew Brzezinski (Presidential Scientific and Senior Advisor throughout the Cold War period) believed then, that weather war was inevitable.[47]

It seems reasonable that the heating and cooling of the upper and mid-atmospheric belts should 'create' artificial pressure zones and 'move' weather patterns; as temperature and pressure dictate weather. (Albeit by massive arrays drawing megawatts of electrical power.)

It's hard to say that the period of anomalous booms over North America between 1977 & 1982 is directly connected to Woodpecker, but the timing is curious. Another curious anomaly was observed between 1980 & 81.

In the February 2nd, 1981, issue of *The Washington Post*, an article states:

> [F]or the past four months, a single weather pattern has gripped the entire United States, causing a coast-to-coast drought unique in the annals of weather recording.[48]

Bearden states:

> [I]t was also one of the most artificial ones ever

---

47. infowars.com
48. The Washington Post. 2.2.1981

recorded! [42]

In regards to Woodpecker, Bearden asserts that it was a Soviet test of America's ability to detect EM scalar waves. He writes that U.S. inaction would indicate their lack of EM weapons understanding.[50]

Of course it is now clear that the U.S. possessed knowledge of scalar weaponization in all areas, so a new idea for Woodpecker's purpose comes to mind: if Woodpecker was a modified OTH radar array on steroids, was its annoying sound and disrupting interference *intended to draw attention* to its signal, with the question being: if the U.S. knew of and could detect Woodpecker's scalar component, where the Soviets testing the ability of *other* nations to spot their scalar game; listening in on, and spying on foreign governments and Intelligence groups to discover 'who knew'?

And further to the (highly speculative) idea of a Cold War U.S./ Soviet alliance, was this a joint U.S./ Soviet scalar test, with full U.S. knowledge?

Note: after Soviet collapse (a possibly staged event) Woodpecker stopped. However less than two years later, larger and vastly more powerful phased array 'installations' suddenly appeared - now known as

49. Bearden. Paper: *Historical Background of Scalar EM Weapons (1990)*
50. Ibid

HAARP arrays after the Alaskan Project (High-frequency Active Auroral Research Project)

By now, many scientists have written about the patents, weaponization, and dangers of HAARP arrays - but the question is: are HAARP arrays, (which now cover the globe), scalar weapons which can't be hidden from sight, so are hidden behind a 'research' label.

If not, why do so many countries need to repeat the same dangerous 'research'? HAARP, as per its early mission statements, 'lifts' sections of the ionosphere for communications, surveillance, and research purposes. The delicate balance of interconnected belts of charged particles that we call the atmosphere, is of course, our only protection against deadly solar and cosmic radiation.

And if so, is weather a weapon?

It may seem strange to allow drought, or the moving south of super-cold flows over ones own country, but in considering this as an 'option', we should further consider:

1) For decades, the CIA allowed 'The Moscow Signal' to be directed at U.S. Embassy staff in Moscow, with full knowledge that it could cause harm. No warnings were issued, nor was any attempt made to stop the intrusion - this leads one to assume that the Russian irradiation of the U.S. Embassy was

allowed to continue purely for the benefit of experimental purposes.[51]

2) Under MK-ULTRA, the CIA laced unsuspecting citizens, soldiers, and colleagues with the hallucinogen LSD, again for experimental purposes.[52]

3) According to the National Cancer Institute (U.S.), nuclear testing in the 1950s & 60s caused at least 75,000 cases of cancer, caused by I^{131}. Those caused by the more sinister Strontium-90, were never tested for. Nuclear fallout was indeed known about.[53]

4) Depleted Uranium rounds, made from spent nuclear fuel rods, have been used by U.S. ground forces in Afghanistan and Iraq with horrific results, that were of course, clearly predictable.

5) 'Collateral Damage' is not just acceptable, it is a staple of U.S. Intelligence and Military doctrine. Hundreds of examples exist.[54]

To inflict drought or storms on its own population, whilst remaining in total denial, is clearly not beyond this régime (whatever it is), especially if gaining a level of control over global weather is a potential outcome.

---

51. wikipedia: The Moscow Signal
52. Maia Szalavitz. A neuroscience journalist for TIME
53. Smith. HAARP: *The Ultimate Weapon of the Conspiracy*
54. Stone. A Time for Deception, pg188

6) If such a régime exists, it clearly operates 'above-government', but also 'in-government'.

Things That Go Bump in the Night

A 1993 patent, (№: *5,038,664*), held by Dr. Bernard Eastlund, is entitled: *Method for producing a shell of relativistic particles at an altitude above the Earth's surface.*

Another Eastlund patent, №: *4,873,928,* Relates to HAARP's ability to produce 'nuclear sized explosions'.

These patents show both a weapons or 'strike' potential, and an ICBM 'shield' potential from today's HAARP arrays; an unthinkable concept outside of science-fiction to laypersons of the 1950s, 60, and 70s.

Colonel Bearden in his, *Fer De Lance,* links Soviet post-war weapons development to the birth of scalar EM weapons; HAARP arrays are simply the apex of that line of research and development.

Throughout September of 1979, (the same month as the anomalous 'Vela Events') a British war cameraman named Nick Downie, reported seeing what he described as 'colossal expanding spheres of light', from deep within the Soviet Union. Downie saw these giant spheres from various vantage points within Afghanistan.[55] These spheres are not mentioned in the available Vela data.

Bearden cites at least eight similar sightings from deep within the Soviet Union by airline pilots in the early 1980s, all describing giant globes of light strangely absent of shockwaves.[56]

On July 20, 1982, the crews of Japan Airlines flights 403 and 421 reported sighting a giant expanding globe of light, 700km east of Kushiro.[57]

Bearden suggests this is likely developmental testing of technology synonymous with Eastlund's later patented HAARP 'Particle shield'. A so named 'Tesla Dome' is in theory, a secret electro-magnetic missile defense shield, capable of protecting large areas from missile and aircraft attacks by duding electronics and exploding ordnance.

But then a more disturbing pattern starts to emerge.

On April 9, 1984, a giant glowing mushroom cloud suddenly rose off Japan's eastern coast. As observed by airline pilots, the cloud rose to 60,000ft and expanded to a width of over 200 miles.[58]

A former military pilot captaining an American Airlines Boeing 747 took evasive action, believing the

---

55. Seitz. Revenge at High Tor
56. Bearden. Paper: *Historical Background of Scalar EM Weapons (1990)*
57. Ibid
58. Ibid

mushroom-cloud to be from a large nuclear explosion. However, the captain was surprised that no shockwave, sound, or blinding flash came.[59]

No explanation for the event was ever given.

Bearden suggests the cloud's source was a Soviet EM howitzer, pulsing in endothermic mode, drawing up a cloud of moisture due to the creation of a low-pressure vacuum;[60] but is there a more sinister explanation?

Off Japan's eastern coast, where this event occurred, the plate boundaries of the North American, and Pacific tectonic plates meet. This mega-fault regularly produces powerful and destructive earthquakes in Japan, and is the fault system that produced the devastating 2011 Fukushima quake (one of the largest earthquakes on record).

Similarly, both Bouvet, and The Prince Edward islands (the sight of the 1979 Vela 'flash') sit *directly* atop the plate boundaries of the mighty African and Antarctic tectonic plates.

If Soviet engineers had indeed developed powerful scalar EM generators (and it certainly appears to be the case) would they naturally be drawn toward the big question, the same question that

---

59. Ibid
60. Ibid

Zbigniew Brzezinski asked: can we trigger earthquakes?

In-fact, Soviet military weapons developers at the time would have been derelict in their duty if they did not explore such an option.

Note that as far back as 1912, Nikola Tesla stated that it was possible to split the planet with the correct resonance, and in 1935 said:

> "[It[becomes possible to convey mechanical effects to the greatest terrestrial distances and produce all kinds of effects - The invention could be used with destructive effect in war."[61]

Both Farrell and Bearden show that post WWII, and in response to U.S. nuclear weapons achievements, Stalin drove his military/ scientific community hard, stating in no uncertain language that the next major technical weapons breakthrough had better be Soviet. A scouring of every and all possible fields of physics to regain superiority (according to Bearden) led the Soviets to Whitaker's 1903 electro-gravitational theories, (theories which are pre-relativistic, but fit with today's advanced theories of quantum mechanics), and also to, Tesla's 'death-ray', now known as a Tesla EM Howitzer.

Bouvet Island as it happens, is one of the

---

61. Ibid

remotest places on Earth - and sitting atop a giant 'fault system' - what better place would there be for testing such an idea as scalar earthquake induction for the first time in human history? The theory being that Extremely Low Frequency (ELF) waves could achieve resonance in the fault-line, causing it to slip, triggering a full fault earthquake. The limiting factors as a weapon would be:

1) Only areas with a fault line could be attacked this way, and;

2) The faults would possibly not give a reliable or consistent response.

If the Bouvet Island incident *was* a Soviet attempt to trigger an earthquake, it failed.

The question is: did their failure push them toward a more 'active' fault zone - one with more strike potential - the giant plate boundary off Japan's eastern coast?

If so the incident of April 9, 1984 did not yield results either.

But then it started.

Tremors In The Deep

In *A Time for Deception,* I noted that Doctors J.P Farrell and Rosie Bertell both cite a strange earthquake in their works; a quake which struck China on July 28, 1976, killing 650,000 people. The odd thing about this event, is that the Chinese military immediately accused the Soviet Union of triggering the quake. Their reasoning was odd for the time: that a Soviet *ionosphereic heater array,* running at full power at the time of the quake, which painted the sky with an un-natural auroral glow, was the quake's cause.[62] (italics added)

Bearden's paper adds:

> [B]efore the first tremor in the early morning, the sky light up like daylight, with multi-hued lights seen up to 200 miles away. Electrical signals were also detected.[63]

Dr. Rosie Bertell cites a range of seismic anomalies

---

62. Stone. A Time for Deception, pg192
63. Bearden. Paper: *Historical Background of Scalar EM Weapons (1990)*

from the late 1980s:

> [On] 12 September 1989, magnetometers at Corralitos
> (near Monterey Bay, in California) detected unusual
> ultra-low frequency waves between 0.01 Hertz and 10
> Hertz. This is the lowest range of ELF waves. These
> waves grew to about 30 times their first intensity, and
> finally subsided on 5 October 1989. On 17 October
> they suddenly appeared again at 2:00pm local time, with
> signals so strong that they went off the scale. Three
> hours later the San Francisco earthquake took place.
> On 29 March 1993, the *Washington Times* reported that
> 'satellites and ground sensors detected mysterious radio
> waves or related electrical and magnetic activity before
> major earthquakes in Southern California during 1986-
> 7, Armenia in 1988, and Japan and Northern California
> in 1989. The 17 January 1994 earthquake in Los
> Angeles was also preceded by unusual radio waves and
> two sonic booms.[64]

Once again, in 2008 after a mega-quake in China's
Sichuan province, the Chinese government made an
accusation of human influence in an act of war: this
time they did not point the finger at Russia - they
accused the United States of killing over 90,000. At
the time of the quake

> [t]he Alaskan HAARP array was running at full power
> and pulsing in the ELF range, peaking before the
> quakes started, and dropping immediately after.[65]

---

64. Bertell. Planet Earth: *The Latest Weapon of War*
65. Stone. A Time for Deception

The Haiti government also believed a U.S. HAARP array triggered its devastating 2010 quake.

The Real Illusion Paradox

At this point there is a serious problem that needs addressing; and that is one of Soviet/ U.S. collusion, despite being engaged in history's mighty 'Cold War'.

Firstly, a massive problem with the existence of Soviet EM weapons and U.S. knowledge of them is this: such EM weapons can function to detonate both conventional explosives, and nuclear warheads, as they stand in place.[66] With the Cold War United States building and holding vast stockpiles of thermonuclear warheads, the Soviets (in theory) could have initiated a nuclear disaster, annihilating the mainland USA. If indeed a brutal and psychopathic archenemy, as we were/are led to believe, what stayed the Soviet hand?

Several reasons are given:

1) That the nuclear fallout from such a strike would cause global radiation poisoning, likely rendering the Soviet Union un-survivable; but EM weapons offer *other*, more practical types of First Strike options and;

---

66. Bearden. Paper: *Historical Background of Scalar EM Weapons (1990)*

2) That another nation, possibly Israel, was equipped with scalar EM weapons early on in the Cold War:

> [O]ne may further speculate that this could reveal what has been checking the Soviets from simply moving against the West with scalar electromagnetic weapons.[67]

But another explanation may exist - a third alternative.

3) What if at the 'highest' level of the pyramid of power, the two 'superpowers' are controlled by the *same* group? This highest level may very well be above the governments of both countries; which appear to be run by their Gestapo like Intelligence arms, and massive, interconnected corporations, forming a massively invested and incestuous global Military Industrial Alliance.

To more fully grasp this (highly speculative) point, and for the purpose of demonstration, we can redraw a *possible,* but much condensed MIC command chart.

---

67. Bearden. Fer De Lance, pg353

US MIC Cmd Level	Level of knowledge / Level of belief in current illusion	Russian Equivalent
Majestic?	Total Allegiance as one MIC Borderless, country-less	Russian Majestic
Cosmic, Luna, Ultra	Secret Space Cold War with Russia Understand that humans invented and operate UFOs, and that No 'aliens' exist or have ever been discovered. Controls scientific suppression of such technologies. Created and manages the deception montage below their level. Black Space Program / EMG propulsion / EM weapons Works to propagate the lower deception plans such as terrorism & Alien presence	Russian Equivalent
Stellar, Astral, Cosmos, Orbit	Continuing 'secret' *terrestrial* 'Cold War' with Russia Aliens fly UFOs / Public Space Program Only / rocket engineering propulsion only / HAARP is a mighty EM	Russian Equivalent

	Weapons platform	
TS-Crypto/ USAP	Continuing secret 'Cold War' Aliens fly UFOs HAARP is a research facility	Russian Equivalent
Top-Secret / Military	US superiority / Relative peace with Russia/ Aliens fly UFOs not humans/ Islamic terrorism is world's biggest threat followed by climate change	Russian Equivalent

This chart will of course, be off the mark in terms of what each 'compartment' knows and believes, or does not know and believe, and the interleaving of each band's knowledge levels - it is of course, simply intended to show how soldiers, generals, FBI agents, FBI directors, politicians, U.S. Presidents, CIA operatives, CIA Station Chiefs, CIA Directors, NSA directors, The Defense Industrial Security Command (DISC), and *Majestic?* Can all believe very *real,* but very different 'illusions' concerning what technology exists, and who the 'enemy' is - and what is conspiracy, and what is real.

This is the 'Real Illusion' paradox.

The real illusion paradox underpins the Cold War deception montage.

In the end, the enemies and technologies associated with each Military/ Industrial Security 'Compartment' are entirely real, but are manipulated by those at the top - the puppet masters of the sickest pantomime ever imagined - drawing trillions into the MIC's coffers via the tax siphon, continual war, and forced global dependence on expensive, obsolete, carcinogenic energy sources.

Effectively, each 'section', made up of two or three 'security compartments', has its own budgets and income streams, its own Military Industrial Complex, its own secret societies, and its own 'continual' wars.

This may sound off-the-mark, but nonetheless, it is where the labyrinth of observable evidence throughout seven decades of deception leads the investigator.

Most (if not all) scholars in this field of research suggest that the 'puppet-masters' of the global pantomime are not connected to any particular country, religion, or for that matter, anything at all; *they* are happy to put humans from whatever creed into their trillion-dollar meat grinder - turning meat into money - The United States of America has been used most, simply because:

1) It has a large, wealthy population and economy.

2) Its people are far more 'patriotic', and therefore easier to manipulate than those of European countries.

3) It 'ain't broke' so don't fix it.

C. Wright Mills, in his 1956 book, *The Power Elite*, stated:

> [W]hen men of knowledge do come to a point of contact with the circles of powerful men, they come not as peers, but as hired men.

The subject of Cold War Soviet/ U.S. complicity is extensive enough to justify an entire book of its own, but several big issues stand out.

1) We mentioned the first 'hot wars' of the Cold War period, Korea and Vietnam, and noted that Soviet Russia simply supplied 'emerging communist states' with conventional weapons in a similar, lucrative system to our own MIC. This then suggests that long, low-tech wars would have also served Soviet interests, as this type of conflict takes from the people, and transfers the tax-take into MIC accounts, and to central bank lenders. (Because governments loan all war funds from a 'central bank', which charges interest, which is covered by taxpayers.)

2) During the early 'space-race', the timing and staggering of Soviet satellite launches and U.S. launches appears staged. Two Soviet launches

followed by three U.S. launches etc. Soviet achievement ahead of the U.S. and then we finally nudge ahead at the final hurdle.

3) Both nations built up nuclear arms and moved them into strategic positions. Televised and radio-program fear campaigns made us understand the horrific danger of nuclear weapons and their use - but nuclear weapons were clearly not the cutting edge WMD - scalar weapons were clearly developed and held by at least the Soviet Union, who did not use them against America. It is totally contradictory to possess nuclear weapons and scalar weapons at the same time. Scalar weapons can produce nuclear sized explosions.

This makes the build-up of nuclear weapons in both countries appear to be a campaign of fear disguised as an 'arms race'. A clever deception to make Soviet Russia and America appear at a hostile stalemate.

4) During the 1978-1989 Afghan conflict a 'secret' CIA supplied Mujahideen was able to defeat this Soviet superpower over ten long years of war - supposedly by financial degradation - several problems with this scenario exist:

(a) The USSR's MIC actually made billions out of this war, whilst reducing much of Russia to poverty.

(b) The USSR possessed both nuclear, and as we've seen, scalar weapons at this time. Oddly however, they chose to let their beloved Union of Soviet Socialist Republics capitulate at the hands of a rag-tag militia, rather than use these weapons.

(c) The 'collapse' of the Soviet Union fits perfectly with the creation of terrorism and the move to the new 'enemy' as stated in the Rosin affidavit. This suggests the Soviet 'collapse' was planned and staged as part of a larger plan.

(d) During the Soviet/ Afghan war of the 1980s, the CIA funded the Afghan side. Anti communist propaganda had long permeated American culture as the brutal 'enemy'. Footage of Soviet nuclear attack saturated TV coverage. Fear was visceral. Yet throughout the 80s, when Russia could not sell their oil, Russian oil was piped from the Caspian region into Iran and to waiting U.S. ships. The Soviet 'enemy's' oil was then taken to America and sold globally as West-Texas Crude.

If financial degradation leading to Soviet capitulation really was the U.S./ CIA goal of the 10 year war in Afghanistan (as stated by Zbigniew Brzezinski) - why did *they* 'help' Russian oil exportation? Why did they assist the enemy? Is it because they were only a 'manufactured enemy'?

In truth, this Soviet/ CIA oil 'deal' had been running since the early 1950s, when the name Stalin was not a good international brand. A company called Occidental Petroleum actually built oil pipelines and infrastructure along the Caspian Sea and into Iran. The deal then appears to be that the CIA brokered the Sale of Stalin's oil, and arranged its sale under the more lucrative West-Texas branding.

Note that this type of dealing actually made Britain more of a 'real enemy' than Soviet Russia during the Afghan War. Russian were not going to strike America, as we were the hand that was feeding them (or some of them). A British/ American 'squabble' to control Iranian oil exports (from 1950-1988) ensued. Both the British and the U.S. used their 'Gestapos' to overthrow and install 'their' men in Iran to gain control of Iranian oil, and this Western meddling of assassinations and heavy-handed control generated an understandable anti-western mindset.

It is an important side note that in 1988, Standard Oil of America and British Petroleum (BP) merged, ending this 'squabble'. BP-America now controls Standard Oil's old Cold War oil subsidiaries.

But for us the fact remains - the American Military Industrial Complex did not view or treat Russia as an enemy, whilst the rest of us were shown image after image of nuclear annihilation.

Russia was *our* enemy - the public's enemy.

Now that enemy is of course, terrorists - ironically a product of the 'Cold War' with Russia.

A Strange DISC

In the above section I mentioned the little known, Defense Industrial Security Command, mischievously 'acronymed' DISC.

DISC goes way back, deep into the Cold War's gestation, and possibly even to the division and establishment of two compartmentally separated space-programs - one based on Werner Von Braun's rocket engineering and a Cold War 'race' against the Soviet Union to the Moon (and space domination); and, the second 'black' space program, using captured Nazi discoid designs, coupled with Tesla's Electro gravitational/ electromagnetic propulsion, producing the first, wobbly 'flying saucers' capable of ultra high-speed, near-space travel.

The latter *never* admitted to - and hidden behind two simultaneous, yet opposite CIA propagated deception plans, which are:

1) UFO's do *not* exist, and;

2) They *do* exist, but are extraterrestrial.[68]

These two lies cover the truth: That UFOs and high-performance discoid spacecraft, from the 1940s to present, are:

1) Of human design and manufacture; with no 'alien back-engineering', and;

2) That intelligent extraterrestrial life has not been found, but the myth of which has been a fantastic cover for U.S. development of Nazi designs and theoretical technology since 'Roswell'.[69]

According to Dr. Carol Rosin (Fairchild Industries Executive during the Cold War), Werner Von Braun, Hitler's number-one rocket engineer, was near the top of the MIC deception pyramid; possibly one of the few with a foot in *both* space-programs - the public program, and the 'black'. Interestingly, Von Braun was the head of DISC, based out of Redstone Arsenal, Alabama.[70]

DISC's role was to ensure the security of MIC associated space industry companies and assets. Even more interestingly, J.P. Farrell draws a connection between the infamous Torbitt Document and the Permanent Index Corporation, a-k-a, Perm-index.

If we cast our minds back to 1963, in his

---

68. Lyne. Pentagon Aliens
69. Stone. A Time for Deception, citing: Lyne, Pentagon Aliens
70. Farrell. The SS Brotherhood of the Bell, pg61

dangerous investigation into government conspiracy and the assassination of John. F. Kennedy, New Orleans District Attorney, Jim Garrison, impeached one Mr. Clay Shaw, on the grounds of conspiracy to commit murder; the same Clay Shaw that sat on the Perm-index board of Directors.

Farrell cites *NASA, Nazis, and JFK: The Torbitt Document:*

> [T]he principle financers of Permindex were a number of U.S. oil companies, H.L Hunt of Dallas, Clint Murchison of Dallas, John DeMenil, Solidarist director of Houston, John Connally as executor of the Sid Richardson Estate, Halliburton Oil Co., Senator Robert Kerr of Okalahoma, Troy Post of Dallas, Lloyd Cobb of New Orleans, Dr. Oechner of New Orleans, George and Herman Brown of Brown & Root, Houston, Attorney Roy M. Cohn, chairman of the board for Lionel Corporation, New York City, Schlemley Industries of New York City, Walter Dornberger, ex-Nazi General, and his company, Bell Aerospace, Pan American World Airways and its subsidiary, Intercontinental Hotel Corporation . . .

> And last but by no means least

> . . . NASA contractors directed by the **Defense Industrial Security Command.** [71] (emphasis added)

The Torbitt Document itself states Perm-index's main

---

71. Farrell. The SS Brotherhood of the Bell, pg62; citing NASA, Nazis, and JFK: *The Torbitt Document*, pgs49 & 50

purpose (at the time of the Kennedy assassination) as:

> 1. [To] fund and direct assassinations of European, Mid-East, and world leaders considered threats to the Western World and to the petroleum interests of the backers.[72]

These 'Big Oil' connections, and their inter-linked relationships with *Paperclip* Nazis, J.F.K's death, and their decades later rise to riches through prolonged Military and Industrial operations in the Middle East, (all predicted in the Rosin affidavit), highlights the existence, seriousness, and brutal nature of the lower-levels of the Military Industrial pyramid - at the time overseen by Von Braun and DISC, who were clearly representing the level or levels above.

Perm-index in today's terms has essentially morphed into the Council on Foreign Relations (CFR), the Military Industrial Alliance tied to the Bilderberg Group, even bigger Oil, Presidential manipulation, and Mideast drone wars.

If we pause for a moment to look at one of the more sinister names on the 1963 Perm-index list, Walter Dornberger, a Nazi General captured and re-employed under *Operation Paperclip* into America's Military Industrial Complex.[73]

Dornberger was part of the Nazi V2 rocket

---

72. NASA, Nazis, and JFK: *The Torbitt Document*, pg49
73. Ibid

program under Von Braun. Whatever the case, Dornberger, who should have been tried at Nuremberg, was elevated to the 'Perm-index level' of the post war MIC, whilst his Nazi superior, Von Braun, went much further.

Secret Space
Wars

Give Me Space

In a recent but seemingly innocuous televised political discussion at a South Korean nuclear summit, believing he was off camera, off air, and in private, President Obama said this to Russian diplomat, Dmitry Medvedev: "On all these issues, but particularly missile defence, this, this can be solved – **but it's important for him to give me space**."

Medvedev nodded, paused, and then said: "I understand. I understand your message about space. Space for you . . . I will transmit this information to Vladimir."[74]

Republican Party spokesmen immediately made a song and dance, referring loudly to 'missile defense' as the source of Obama's need for 'space'. There appears to be nothing odd in this exchange in our current paradigm of thinking, but it is a conversation that I will return to soon.

---

74. *www.theguardian.com/world/2012/mar/26/obama-medvedev-space-nuclear*

A Bolt from the Blue: STS-48

If you have not yet seen it, view the NASA space shuttle 'STS-48' mission clip on you-tube. The film was taken in mid September of 1991 from the space shuttle *Discovery,* which during the STS-48 mission, completed 81 revolutions of Earth.

Sunlight reflecting off numerous ice particles is clearly obvious in the video footage and is totally normal, but whilst above the South Pacific, Discovery's cameras recorded what appears to be a UFO under intelligent control. Initially the dot of light appears as the other space-debris - a dot of light in the distance, slowly approaching Earth. Then out-of-the-blue, a bolt of 'lightning', (possibly from an Earth based weapon), shoots directly at the 'dot' from the planet's surface. The result is stunning: the 'ice-particle' makes a sudden evasive turn of incredible speed and accelerates away - the plasma bolt narrowly missing it.[75]

---

75. NASA STS-48, youtube

Ice particles for the record, do not turn and accelerate.

NASA officially explains the incident as "ice particles reacting to jet engines". Ufologists on the other hand cite the STS-48 footage as clear proof of extraterrestrial life - and a possible alien 'space war'. It is however, far more reasonable to assume the spacecraft to be of Earthly origin, representing U.S. Russian, or Chinese top technology. It is also reasonable to assume the 'plasma bolt' to have come from one of the many HAARP arrays, with interleaving arcs of fire covering Earth's entire surface; a machine of course, which is designed to do exactly that - 'shoot' one-billion watt beams into the upper-atmosphere.

If HAARP arrays are firing on 'enemy spacecraft', it certainly shows the existence of a secret 'space war' - but a war that is far more likely one between *human* superpowers.

So why should we assume a solely human origin to this UFO phenomenon? Firstly let's look at what we know:

1) Late in WWII, Nazi engineers had designed several disk-shaped flying-craft known as Haunebu.[76] The single explanation for a discoid shape is the invention of a very different type of propulsion - a propulsion

---

76. Farrell. The SS Brotherhood of the Bell

system that does not use aerodynamics - possibly Tesla's theoretical electro-dynamic propulsion system.[77]

2) Sightings of wobbly flying saucers literally 'took-off' in the late 1940s, after such technology was recovered by U.S. technology sweeps of Nazi WWII European zones of operation primarily under projects *Lusty,* and *Paperclip.*

3) These vintage 1940s-50s era disks (complete with portholes and rivet seams no less) were indicative of the time, and also of their early stage of development. They had clearly not flown for many years across interstellar space at a required velocity approaching light-speed.

4) The quality of UFO design has improved apace with human advances in technology. (The portholes are gone.)

5) The Roswell 'crash' was a clear piece of CIA deception, rapidly orchestrated to propagate the myth of extraterrestrial visitation, which predictably under deception doctrine, served to cover the development and testing of *our* discoid craft with their obviously secret propulsion system.[78]

---

77. Lyne. Pentagon Aliens
78. Ibid

For more on the Roswell deception, see *A Time for Deception;* here is an excerpt:

[I]n his book *Pentagon Aliens,* Lyne asserts that not one, but two, Roswell ufo crashes were staged; the first involving a man-made aluminum discoid 12ft in diameter with a bubble canopy and an automobile aerial, complete with three dead rhesus monkeys, shaved, died green, and dressed in G suits. (These monkeys and their g-suits were employed in the rocket-sled g-force testing facility at the nearby Alamogordo AFB at the time and actually had these types of g-suits for their experiments).

Lyne asserts that it became obvious that the staging work at the initial 'ufo crash site' was below par, so it was cleared up and aluminum foil and weather balloon parts were introduced.

Lyne points out several important things:

Discoid craft, either German or those built from Nazi blueprints, could not be used in the Roswell deception plan because of the sensitive 'National Security nature' of this cutting edge technology, and;

Intelligence personal involved were prohibited from disclosing any details due to 'National Security', and the Classified Information Act. (And possibly some subtle coercion).

Lyne sites physical problems with the Roswell crash scene like; the furrow in the hard stony desert being too deep for the mass and inertia of the tiny craft, and;

The use of exaggerated, mis-stated, overstated, and false witness reports; and the use of CIA and/or DIA 'planted' witnesses'.

Lyne's message on the Roswell deception is this: the two big simultaneous and opposite lies that the highest compartment of the Intelligence machine push to maintain their Roswell ET origin deception, are:

1. That flying saucers do not exist; and,

2. That they do exist and are extraterrestrial.

Lyne also asserts that: the most likely alternative is:

3. That all ufos are "real, man-made flying machines, invented and flown by human beings for over fifty years, there being no alien contacts whatsoever on Earth."[79]

6) The problem of time dilation (the slowing of time for those traveling at very high speed for long periods, as would be required in order to travel from even our closest neighboring star system) would make it highly problematic for 'aliens' to travel across space to Earth, for they would not be able to return home within their own generational lifetimes.

[On] modern (and I use the term loosely) battle-fields drones do the work of tactical reconnaissance and Close-Air-Support (CAS) aircraft - their pilots safe back in the U.S, can even sleep in there own beds at the end of each mission.

It makes sense that if we have built spacecraft capable of interstellar flight, that we should pilot them remotely - drone spaceships.

The Drone Paradox, like the Twins Paradox, is not

79. Stone. A Time for Deception

really a paradox at all, but simple fact based on the laws of Relativistic physics.

So if a space-drone fitted with a forward looking camera was flown by pilot based in a dark room deep under the Pentagon at 99% of light-speed directly toward the star Sirius A, the drone would take around 9 years to arrive - however the pilot back in the Pentagon would be dead and buried long before seeing the drone's arrival in the Canis System, because it would take 135 years of Earthly time at the pilots computer consol.

Do this at 90% C however, and the figures become more realistic, but still challenging.

For humans traveling off-planet, a 1 year return flight from Earth at 0.99C would have the craft arriving back about 15 years after its day of departure. (But only 1 year's elapsed time on the clock inside the craft). This is because the faster one travels, the slower time effects him relative to time on Earth.

At full light-speed, it is theorized that time stops altogether - it is also said that matter cannot reach light-speed, but only a percentage of it.

About 2000 stars sit within a 50 light-year radius from Earth. As we saw earlier, if an ET discoid left its home plant sitting 40 light-years from Earth and traveled at near light speed, by the time of their arrival, about 600 years would have passed on their home plant.

This makes interstellar flight possible but very problematic. It seems that even at the outer reaches of possibility, less than 2000 star systems are within possible traveling distance from us here on Earth.

Conversely, if a human craft left Earth for say Saturn's moon, Titan, and accelerated instantly to 90% light-speed and the trip took 10 minutes - an Earth bound observer would only have to wait 22min and 54sec to witness its arrival on Titan. (Assuming the 1min: 2.29min ratio for 90%C.)[80]

7) Decades of deception have been created by the CIA around UFO activity for one logical reason; it is *our* technology, and for reasons of plausible deniability and total dominance, it is better that everyone believes it's 'aliens'.

8) The end of WWII is clearly the point at which the two space programs were established. The Von Braun/ NASA rocket space program reinforces our belief that 'discoid craft' are clearly 'alien'. The two space program ruse adds another, deeper level of deception to the already complex Cold War deception montage.

These basic points suggest a solely human nature to the UFO phenomena.

A Strange But Very Possible Speculation

Remember that Obama thought no one could hear him when he said to Medvedev: "On all these issues, but particularly missile defence, this, this can be solved – **but its important for him to give me**

---

80. Stone. A Time for Deception

space." Obama used an unusually animated body language, including a strange eye engagement, as if suggesting a furtive meaning to his statement, and that Medvedev should 'read between the lines'.

Medvedev nodded, paused in animation, and then (if Obama's words did not convey secret meaning) gave a quite unusual, and unusually animated answer: "I understand . . . I understand your message about space . . . Space for you . . . I will transmit this information to Vladimir. I understand."

So, we have discoid flying spacecraft; ufos are a human invention, first envisioned by Nazi engineers tapping the genius of Nikola Tesla, and then won by the United States during WWII's prolonged death later to be covered in elaborate layers of deception. We have Soviet EM weapons tests. We have HAARP arrays covering the globe. We have NASA footage of a UFO clearly under intelligent control, taking evasive action to avoid a bolt of 'plasma' fired from Earth.

In this context, could Obama's message to Medvedev carry a more 'cryptic' meaning?

Are we, like mesmerized fools, watching the magician's hand, luring our eyes and our fears toward the Middle East and the threat of terrorism, whilst the real maneuvering is conducted out of sight in the forms of technology suppression, EM scalar weapons, anti-gravity propulsion, and a 'hot' electromagnetic war in our own solar system?

Seeing Is Believing: Orbs & UFO's

UFO sightings, from the 1940s until present, (although usually lumped into one class - aliens) actually fall into several very different categories.

(1) Flying discs

(2) Flying triangles

(3) Flying cigar shaped craft

(4) Undersea flying UFOs

(5) Tiny UFOs

(6) Giant Ufos

(7) Lights, and;

(8) *Orbs*

Categories 1 thru 7 can be explained as various forms of human engineered secret strike aircraft, surveillance aircraft, hovering or flying secret drones, stealth aircraft, and ultimately, electro-dynamic space-going craft.

But what of Orbs?

Orbs are mostly described by witnesses as 'balls of light', 'balls of energy', or 'balls of plasma' that often hover in silence, move slowly, accelerate at high-speed, and mysteriously 'vanish' into thin air.

Orbs don't appear big enough to carry passengers. Orb sightings are highly prevalent, yet totally different to actual 'nuts-and-bolts' space-ships.

Recently the History Channel has aired historical (Cold War era) stories of U.S. fighter aircraft being scrambled to investigate strange glowing orbs in American skies, only for pilots to pursue see theses orbs and see them vanish into thin air.

In 1980 near RAF Bentwaters, a military facility in East England, orbs landed in a nearby forest on two consecutive nights, causing damage to trees and leaving strange marks in the clay.

The base commander asserts because of the orb's or orbs' movements and reactions, that they were under intelligent control.

Bentwaters skeptics suggest that highly experienced air-force officers were fooled by car headlights from the A12 motorway, or even a nearby lighthouse, both of which are refuted by the Bentwaters officers - but is there another explanation?

Colonel Tom Bearden suggests that the glowing and vanishing orbs of plasma being seen across the globe (in unprecedented numbers during the Cold War) are connected to Soviet EM Scalar weapons testing, and calibration.

Bearden believes these orbs are a phenomena of scalar interferometry - possibly visual or detectable targeting markers.

It stands to reason that if one could produce an 'orb' from a computer consol and fly it around the globe, one could clearly have quite a lot of fun.

Bearden asserts that the reason these 'alien' orbs vanish into thin air, is simply because, by the click of a mouse, they have been turned off.[81]

With an increase in the number of HAARP stations and their use, it is reasonable to expect a possible increase in global orbs sightings.

When one considers the science behind scalar interferometry and its reason for further weaponization and development, it is also possible that these HAARP induced plasma orbs could be used in surgical strikes of virtually any kind, such as to bring down aircraft, eliminate a specific ground target, or to fly orbs around to enhance UFO hysteria - and even make crop circles.

---

81. cheniere.org

Why Iraq is Not a Scalar War-Zone

So on the question of: will scalar EM weapons be used in the Middle East? The answer is . . .

Only if necessary.

The Mideast occupation is not designed to be a scalar war, but scalar weapons can be used in countless ways and forms to effect very small strikes with a high level of deniability.

The Mideast situation (as shown by the non-use of decisive EM intervention) is an occupation that has to *appear* to be 'difficult' to contain. The supply and demand of low-tech weapons like bullets and bombs creates a solid income stream in a purposefully prolonged 'conflict', but such weapons also help create the illusion of struggle, re-enforcing and justifying the prolonged nature of Mideast operations.

Scalar EM weapons could clearly be used to halt IS militants with little collateral damage, but they are not used for the simple reason that the CIA wants and needs IS. Terrorism and its publicly perceived difficulty to control, keeps the CIA and its military/ industrial complex in business.

Another point is: if Putin's Russia possesses scalar weapons and electromagnetic defensive shields - as evidenced by the many observations of the testing and use of such machines - then why has Russia never used these weapons to protect its allies in the Mideast region?

Earlier we mentioned Israel's possible EM potential as a deterrent - but when all the evidence from decades of deception is considered, from Cold War complicity until present, it appears more-and-more obvious that at the highest level, Russian and America are one and the same, and the wars of the various military/ industrial complexes below are just highly profitable diversions from a much darker reality.

A Even More Bizarre Speculation

Secret spacecraft, secret beam weapons, a labyrinth of technology deception, and the 'terrorist diversion' as laid out in the Rosin Affidavit - suggests a bizarre and un-comfortable possibility . . .

That a section of humanity, albeit the very elite pinnacle of America's Military/ Industrial pyramid - has a reliable and functioning 'off-planet' capability - and has had for some years.

If all that I have suggested is correct - (and the weight of un-picked deceptions and historical evidence suggests so) - then the human race has a breakaway empire, separate from the constraints of our 'manufactured' low-tech reality, living in an entirely different realm, and possessing an entirely different future.

References

1. Quinion. (2009-04-18). "Afpak". World Wide Words. Af-Pak combines the nations of Afghanistan & Pakistan into one 'war-zone', for U.S. purposes.

2. Stone. ATime for Deception.

3. Stone. A Time for Deception.

4. Ibid.

5. Stone. A Blurred Reality. pg80

6. Holocaust Encyclopedia: Martin Niemöller (1892–1984) was a prominent Protestant pastor who emerged as an outspoken public foe of Adolf Hitler and spent the last seven years of Nazi rule in concentration camps.

7. Marrs. Rise of the Fourth Reich

8. United States Patent №: *4,873,928*. Issued to Bernard Eastlund (at the time a consultant to ARCO industries).

9. Turse & Engelhardt. Terminator Planet

10. http://www.defense.gov/transcripts/transcript.aspx?transcriptid=674

11. Bearden. Fer De Lance

12. The Affidavit of Dr. C. Rosin. Taken from the Disclosure Project website.

13. Soviet capitulation was finalized in 1991

14. Lyne. Pentagon Aliens

15. Ibid

16. Bearden. Paper: *Historical Background of Scalar EM Weapons (1990)*

17. Farrell. The SS Brotherhood of the Bell, pg229

18. Ibid. Farell infers that the development of such an ICBM suggets it was intended to be coupled to an atomic warhead of sorts; for what other reason would there be to go to such lenghts when clearly a conventional warhead would be pointless, as was the case with the V1 and V2 attacks on Greater London.

19. Ibid

20. Farrell. The SS Brotherhood of the Bell, pg231

21. Ibid, pg240

22. chenier.org

23. Beardon. Fer De Lance
24. Ibid
25. (A possible idiom) It is said that that Winston Churchill (with prior knowledge of the bomber attack) let the English town of Coventry fall to German bombers, without warning, to protect his intelligence gathering aparatus prior to D-Day. Some say this is false. Some say it makes sense. Some still say - we've put him in 'Coventry'.
26. http://heasarc.gsfc.nasa.gov/docs/heasarc/missions
27. U.S. State Department: Limited Test Ban Treaty pdf
28. Ibid
29. 1963 Speech on The Limited Test Ban Treaty between the Soviet Union and the United States of America.
30. U.S. State Department: Limited Test Ban Treaty pdf
31. Lewiston Morning Tribune, 1979
32. Ibid
33. Bearden. Fer De Lance
34. Vela Data
35. Ibid
36. Bearden. Fer De Lance
37. CIA factbook
38. Bearden. Paper: *Historical Background of Scalar EM Weapons (1990)*
39. wikipedia. Russian Woodpecker
40. Ibid
41. Ibid
42. Ibid
43. Ibid
44. Bearden. Paper: *Historical Background of Scalar EM Weapons (1990)*
45. Ibid
46. Ibid
47. infowars.com
48. The Washington Post. 2.2.1981
49. Bearden. Paper: *Historical Background of Scalar EM Weapons (1990)*
50. Ibid
51. wikipedia: The Moscow Signal
52. Maia Szalavitz. A neuroscience journalist for TIME
53. Smith. HAARP: *The Ultimate Weapon of the Conspiracy*
54. Stone. A Time for Deception, pg188

55. Seitz. Revenge at High Tor
56. Bearden. Paper: *Historical Background of Scalar EM Weapons (1990)*
57. Ibid
58. Ibid
59. Ibid
60. Ibid
61. Ibid
62. Stone. A Time for Deception, pg192
63. Bearden. Paper: *Historical Background of Scalar EM Weapons (1990)*
64. Bertell. Planet Earth: *The Latest Weapon of War*
65. Stone. A Time for Deception
66. Bearden. Paper: *Historical Background of Scalar EM Weapons (1990)*
67. Bearden. Fer De Lance, pg353
68. Lyne. Pentagon Aliens
69. Stone. A Time for Deception, citing: Lyne, Pentagon Aliens
70. Farrell. The SS Brotherhood of the Bell, pg61
71. Farrell. The SS Brotherhood of the Bell, pg62; citing NASA, Nazis, and JFK: *The Torbitt Document*, pg49 & 50
72. NASA, Nazis, and JFK: *The Torbitt Document*, pg49
73. Ibid
74. *www.theguardian.com/world/2012/mar/26/obama-medvedev-space-nuclear*
75. NASA STS-48, youtube
76. Farrell. The SS Brotherhood of the Bell
77. Lyne. Pentagon Aliens
78. Ibid
79. Stone. A Time for Deception
80. Ibid
81. cheniere.org

Annex I

Annex I

The Operation Northwoods Briefing Document

THE JOINT CHIEFS OF STAFF
WASHINGTON D.C.

13 March 1962

MEMORANDUM FOR THE SECRETARY OF DEFENSE

Subject: Justification for US Military Intervention in Cuba (TS)

1. The Joint Chiefs of Staff have considered the attached Memorandum for the Chief of Operations, Cuba Project, which responds to a request of that office for brief but precise description of pretexts which would provide justification for US military intervention in Cuba.

2. The Joint Chiefs of Staff recommend that the proposed memorandum be forwarded as a preliminary submission suitable for planning purposes. It is assumed that there will be similar submissions from other agencies and that these inputs will be used as a basis for developing a time-phased plan. Individual projects can then be considered on a case-by-case basis.

3. Furthur, it is assumed that a single agency will be given the primary responsiblity for developing military and para-military aspects of the basic plan. It is recommended that this responsiblity for both overt and covert military operations be assigned the Joint Chiefs of Staff.

~~TOP SECRET SPECIAL HANDLING NOFORN~~

THE JOINT CHIEFS OF STAFF
WASHINGTON 25, D.C.

UNCLASSIFIED

13 March 1962

MEMORANDUM FOR THE SECRETARY OF DEFENSE

Subject: Justification for US Military Intervention in Cuba (TS)

1. The Joint Chiefs of Staff have considered the attached Memorandum for the Chief of Operations, Cuba Project, which responds to a request of that office for brief but precise description of pretexts which would provide justification for US military intervention in Cuba.

2. The Joint Chiefs of Staff recommend that the proposed memorandum be forwarded as a preliminary submission suitable for planning purposes. It is assumed that there will be similar submissions from other agencies and that these inputs will be used as a basis for developing a time-phased plan. Individual projects can then be considered on a case-by-case basis.

3. Further, it is assumed that a single agency will be given the primary responsibility for developing military and para-military aspects of the basic plan. It is recommended that this responsibility for both overt and covert military operations be assigned the Joint Chiefs of Staff.

For the Joint Chiefs of Staff:

SYSTEMATICALLY REVIEWED
BY JCS ON ___21 May 84___
CLASSIFICATION CONTINUED

L. L. LEMNITZER
Chairman
Joint Chiefs of Staff

1 Enclosure
Memo for Chief of Operations, Cuba Project

EXCLUDED FROM GDS

EXCLUDED FROM AUTOMATIC
REGRADING: DOD DIR 5200.10
DOES NOT APPLY

~~TOP SECRET SPECIAL HANDLING NOFORN~~

Pg 1 of the Northwoods document

JUSTIFICATION FOR MILITARY INTERVENTION IN CUBA

THE PROBLEM

1. As requested by Chief of Operations, Cuba Project, the Joint Chiefs of Staff are to indicate brief but precise description of pretext which they consider would provide justification for US miltary intervention in Cuba.

FACTS BEARING ON THE PROBLEM

2. It is recognised that any action which becomes pretext for US military intervention in Cuba will lead to a political decision which then would lead to military action.

3. Cognizance has been taken of a suggested course of action proposed by the US Navy relating to generated instances in the Guantanamo area.

4. For additional facts see enclosure B.

DISCUSSION

5. The suggested course of action appended to Enclosure A are based on the premis that US military intervention will result from a period of heightened US-Cuban tensions which place the United States in the position of suffering justifiable grievences. World opinion, and the United Nations forum should be favorably affected by developing the inter-national image of the Cuban government as rash and irresponsible, and as an alarming and unpredictable threat to the peace of the Western Hemisphere.

6. While the foregoing premis can be utilised at the present time it will continue to hold good only as long as there can be resonable certainty that US military intervention in Cuba would not directly involve the Soviet Union, there is

---

Operation Northwoods

UNCLASSIFIED

JUSTIFICATION FOR US MILITARY INTERVENTION IN CUBA (TS)

THE PROBLEM

1. As requested[*] by Chief of Operations, Cuba Project, the Joint Chiefs of Staff are to indicate brief but precise description of pretexts which they consider would provide justification for US military intervention in Cuba.

FACTS BEARING ON THE PROBLEM

2. It is recognized that any action which becomes pretext for US military intervention in Cuba will lead to a political decision which then would lead to military action.

3. Cognizance has been taken of a suggested course of action proposed[**] by the US Navy relating to generated instances in the Guantanamo area.

4. For additional facts see Enclosure B.

DISCUSSION

5. The suggested courses of action appended to Enclosure A are based on the premise that US military intervention will result from a period of heightened US-Cuban tensions which place the United States in the position of suffering justifiable grievances. World opinion, and the United Nations forum should be favorably affected by developing the international image of the Cuban government as rash and irresponsible, and as an alarming and unpredictable threat to the peace of the Western Hemisphere.

6. While the foregoing premise can be utilised at the present time it will continue to hold good only as long as there can be reasonable certainty that US military intervention in Cuba would not directly involve the Soviet Union. There is

---

[*] Memorandum for General Craig from Chief of Operations, Cuba Project, subject: "Operation MONGOOSE", dated 5 March 1962, on file in General Craig's office.
[**] Memorandum for the Chairman, Joint Chiefs of Staff, from Chief of Naval Operations, subject: "Instances to Provoke Military Actions in Cuba (TS)", dated 8 March 1962, on file in General Craig's office.

2

UNCLASSIFIED

<del>TOP SECRET — SPECIAL HANDLING — NOFORN</del>

as yet no bilateral mutual support agreement binding the USSR to the defense of Cuba, Cuba has not yet become a member of the Warsaw Pact, nor have the Soviets established Soviet bases in Cuba in the pattern of US bases in Western Europe. Therefore, since time appears to be an important factor in resolution of the Cuba problem, all projects are suggested within the time frame of the next few months.

CONCLUSION

7. The suggested course of action appended to Enclosure A satisfactorily respond to the statement of problems. However these suggestions should be forwarded as a preliminary submission suitable for planning purposes, and together with similar inputs from other agencies, provide a basis for development of a single intergrated, time-phased plan to focus all efforts on the objective of justification for US military intervention in Cuba.

RECOMMENDATIONS

8. It is recommended that:

a) Enclosure A together with its attachments should be forwarded to the Secretary of Defense for approval and transmittal to the Chief of Operations, Cuba Project.

b) This paper NOT be forwarded to commanders of unifed or specified commands.

c) This paper Not be forwarded to US officers assigned to NATO activities.

d) This paper NOT be forwarded to the Chairman, US Delegation, United Nations Military Staff Comittee.

UNCLASSIFIED

as yet no bilateral mutual support agreement binding the USSR to the defense of Cuba, Cuba has not yet become a member of the Warsaw Pact, nor have the Soviets established Soviet bases in Cuba in the pattern of US bases in Western Europe. Therefore, since time appears to be an important factor in resolution of the Cuba problem, all projects are suggested within the time frame of the next few months.

CONCLUSION

7. The suggested courses of action appended to Enclosure A satisfactorily respond to the statement of the problem. However, these suggestions should be forwarded as a preliminary submission suitable for planning purposes, and together with similar inputs from other agencies, provide a basis for development of a single, integrated, time-phased plan to focus all efforts on the objective of justification for US military intervention in Cuba.

RECOMMENDATIONS

8. It is recommended that:

a. Enclosure A together with its attachments should be forwarded to the Secretary of Defense for approval and transmittal to the Chief of Operations, Cuba Project.

b. This paper be NOT be forwarded to commanders of unified or specified commands.

c. This paper NOT be forwarded to US officers assigned to NATO activities.

d. This paper NOT be forwarded to the Chairman, US Delegation, United Nations Military Staff Committee.

3

UNCLASSIFIED

Annex I

. . . Such a plan would enable a logical build-up of incidents to be combined with other seemingly unrelated events to camouflage the ultimate objective and create the necessary impression of Cuban rashness and irresponsiblity on a large scale, directed at other countries as well as the United States. The plan would also properly intergrate and time phase the courses of action to be pursued. The desired resultant from the execution of this plan would be to place the United States in the apparent position of suffering defendable grievances from a rash and irresponsible government of Cuba and to develope an international image of a Cuban threat to peace in the Western Hemisphere.

Operation Northwoods

SPECIAL HANDLING NOFORN

APPENDIX TO ENCLOSURE A

DRAFT

MEMORANDUM FOR CHIEF OF OPERATIONS, CUBA PROJECT

Subject: Justification for US Military Intervention in Cuba (TS)

1. Reference is made to memorandum from Chief of Operations, Cuba Project, for General Craig, subject: "Operation MONGOOSE", dated 5 March 1962, which requested brief but precise description of pretexts which the Joint Chiefs of Staff consider would provide justification for US military intervention in Cuba.

2. The projects listed in the enclosure hereto are forwarded as a preliminary submission suitable for planning purposes. It is assumed that there will be similar submissions from other agencies and that these inputs will be used as a basis for developing a time-phased plan. The individual projects can then be considered on a case-by-case basis.

3. This plan, incorporating projects selected from the attached suggestions, or from other sources, should be developed to focus all efforts on a specific ultimate objective which would provide adequate justification for US military intervention. Such a plan would enable a logical build-up of incidents to be combined with other seemingly unrelated events to camouflage the ultimate objective and create the necessary impression of Cuban rashness and irresponsibility on a large scale, directed at other countries as well as the United States. The plan would also properly integrate and time phase the courses of action to be pursued. The desired resultant from the execution of this plan would be to place the United States in the apparent position of suffering defensible grievances from a rash and irresponsible government of Cuba and to develop an international image of a Cuban threat to peace in the Western Hemisphere.

UNCLASSIFIED   5   Appendix to
                                                Enclosure A

TOP SECRET   SPECIAL HANDLING   NOFORN

viii

Annex I

ANNEX TO APPENDIX TO ENCLOSURE A

PRETEXTS TO JUSTIFY US MILITARY INTERVENTION IN CUBA

- -

1. Since it would not seem desirable to use legitimate provocation as the basis for US military intervention in Cuba a cover and deception plan, to include requisite preliminary action such as has been developed in response to task 33, could be executed as an initial effort to provoke Cuban res**tions. Harassment plus deceptive actions to convince the Cubans of imminent invasion would be emphasised. Our military posture throughout execution of the plan will allow rapid change from exercise to intervention if Cuban response justifies.

2. A series of well coordinated incidents will be planned to take place in and around Guantanamo to give genuine apperance of being done by hostile Cuban forces.

a. Incidents to establish a credible attack (not in chornological order):

(1) Start rumors (many). Use clandestine radio.

(2) Land friendly Cubans in uniform "over-the-fence" to stage attack on base.

(3) Capture Cuban (friendly) saboteure inside the base.

(4) Start riots near the base main gate (friendly Cubans).

Operation Northwoods

 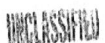

ANNEX TO APPENDIX TO ENCLOSURE A

PRETEXTS TO JUSTIFY US MILITARY INTERVENTION IN CUBA

(Note: The courses of action which follow are a preliminary submission suitable only for planning purposes. They are arranged neither chronologically nor in ascending order. Together with similar inputs from other agencies, they are intended to provide a point of departure for the development of a single, integrated, time-phased plan. Such a plan would permit the evaluation of individual projects within the context of cumulative, correlated actions designed to lead inexorably to the objective of adequate justification for US military intervention in Cuba).

1. Since it would seem desirable to use legitimate provocation as the basis for US military intervention in Cuba a cover and deception plan, to include requisite preliminary actions such as has been developed in response to Task 33 c, could be executed as an initial effort to provoke Cuban reactions. Harassment plus deceptive actions to convince the Cubans of imminent invasion would be emphasized. Our military posture throughout execution of the plan will allow a rapid change from exercise to intervention if Cuban response justifies.

2. A series of well coordinated incidents will be planned to take place in and around Guantanamo to give genuine appearance of being done by hostile Cuban forces.

    a. Incidents to establish a credible attack (not in chronological order):

        (1) Start rumors (many). Use clandestine radio.

        (2) Land friendly Cubans in uniform "over-the-fence" to stage attack on base.

        (3) Capture Cuban (friendly) saboteurs inside the base.

        (4) Start riots near the base main gate (friendly Cubans).

7

Annex to Appendix
to Enclosure A

X

---

(5) Blow up ammunition inside the base; start fires.

(6) Burn aircraft on air base (sabotage).

(7) Lob mortar shells from outside of base into base.

Some damage to installations.

(8) Capture assult teams approaching from sea or vicinity of Guantanamo City.

(9) Capture militia group which storms the base.

(10) Sabotage ship in harbor; large fires - - nepthalene.

(11) Sink ship near harbor entrance. Conduct funerals for mock victims (may be lieu of (10)).

      b. United States would respond by executing offensive operations to secure water and power supplies, destroying artillery and mortar emplacements which threaten the base.

      c. Commence large scale United States military operations.

3. A "Remember the Maine" incident could be arranged in several forms:

      a. **We could blow up a US ship in Guantanamo Bay and blame Cuba.**

      b. **We could blow up a drone (unmanned) vessel anywhere in the Cuban waters.** We could arrange to cause such incident in the vicinity of Havana or Santiago as a spectacular result of Cuban attack from the air or sea, or both. The presence of Cuban planes or ships merely investigating the intent of the vessel could be fairly compelling evidence that the ship was taken under attack. The nearness to Havana or Santiago would add crediblity especially to those people that might have heard the blast or have seen the fire. The US could follow up with an air/sea rescue operation covered by US fighters to **"evacuate" remaining members of the non-existent crew. Casualty lists in US newspapres would cause a helpful wave of national indignation.**

4. **We could develope a Communist Cuban terror campaign in the Maiami area, in other Florida cities and even in Washington.**

(Bolding added)

Operation Northwoods

APPENDIX TO ENCLOSURE A

DRAFT

UNCLASSIFIED

MEMORANDUM FOR CHIEF OF OPERATIONS, CUBA PROJECT

Subject: Justification for US Military Intervention in Cuba (TS)

1. Reference is made to memorandum from Chief of Operations, Cuba Project, for General Craig, subject: "Operation MONGOOSE", dated 5 March 1962, which requested brief but precise description of pretexts which the Joint Chiefs of Staff consider would provide justification for US military intervention in Cuba.

2. The projects listed in the enclosure hereto are forwarded as a preliminary submission suitable for planning purposes. It is assumed that there will be similar submissions from other agencies and that these inputs will be used as a basis for developing a time-phased plan. The individual projects can then be considered on a case-by-case basic.

3. This plan, incorporating projects selected from the attached suggestions, or from other sources, should be developed to focus all efforts on a specific ultimate objective which would provide adequate justification for US military intervention. Such a plan would enable a logical build-up of incidents to be combined with other seemingly unrelated events to camouflage the ultimate objective and create the necessary impression of Cuban rashness and irresponsibility on a large scale, directed at other countries as well as the United States. The plan would also properly integrate and time phase the courses of action to be pursued. The desired resultant from the execution of this plan would be to place the United States in the apparent position of suffering defensible grievances from a rash and irresponsible government of Cuba and to develop an international image of a Cuban threat to peace in the Western Hemisphere.

UNCLASSIFIED

5

Appendix to
Enclosure A

xii

The terror campaign could be pointed at Cuban refugees seeking haven in the United States. We could sink a boatload of Cubans enroute to Florida (real or simulated). We could also foster attempts on lives of Cuban refugees in the United States even to the extent of wounding in instances to be widely publicized. Exploding a few plastio bombs in carefully chosen spots, the arrest of Cuban agents and the release of prepared documents substantiating Cuban involvment also would be helpful in projecting the idea of an irresponsible government.

6. Use of MIG type aircraft by US pilots could provide additional provocation. Harassment of civil air, attacks on surface shipping and destruction of US military drone aircraft by MIG type planes would be useful as complementary actions. An F-86 properly painted would convince air passengers that they saw a Cuban MIG, especially if the pilot of the transport were to announce such fact. The primary drawback to this suggestion appears to be the security risk inherent in obtaining or modify-ing an aircraft. However, reasonable copies of the MIG could be produced from US resources in about three months.

---

7. Hijacking attempts against civil air and surface craft should appear to continue as harassing measures condoned by the government of Cuba. Concurrently, genuine defections of Cuban civil and military air and surface craft should be encouraged.

8. **It is possible to create an incident which will demonstrate convincingly that a Cuban aircraft has attacked and shot down a chartared civil airliner enroute from the United States to Jamaica, Guantemala, Panama or Venezuela. The destination would be chosen only to cause the flight plan route to cross Cuba. The passengers could be a group of persons with a common interest to support chartering a non-scheduled flight.**

a. An aircraft at Eglin AFB would be painted and numbered as an exact duplicate for a civil registered aircraft belonging to a CIA proprietary organization in the Maiami area. At a designated time the duplicate would be substituted for the actual civil aircraft and would be loaded with the selected passengers, all boarded under carefully prepared aliases. **The actual registered aircraft would be converted to a drone.**

　　b. Take off time of the drone aircraft and the actual aircraft will be scheduled to allow a rendezvous south of Florida. From the rendezvous point the passenger-carrying aircraft will descend to minimum altitude and go directly into an auxiliary field at Eglin AFB where arrangements will have been made to evacuate the passengers and return the aircraft to its original status. The drone aircraft meanwhile will continue to fly the filed flight plan. When over Cuba the drone will being[sic] transmitting on the international distress frequency a "MAY DAY" message stating he is under attack by Cuban MIG aircraft. The transmission will be interrupted by destruction of the aircraft which will be triggered by radio signal. **This will allow ICAO radio stations in the Western Hemisphere to tell the US what has happened to the aircraft instead of the US trying to "sell" the incident.**

―――――――――――――――――――――――

(Bolding added)

Annex I

7. Hijacking attempts against civil air and surface craft should appear to continue as harassing measures condoned by the government of Cuba. Concurrently, genuine defections of Cuban civil and military air and surface craft should be encouraged.

8. It is possible to create an incident which will demonstrate convincingly that a Cuban aircraft has attacked and shot down a chartered civil airliner enroute from the United States to Jamaica, Guatemala, Panama or Venezuela. The destination would be chosen only to cause the flight plan route to cross Cuba. The passengers could be a group of college students off on a holiday or any grouping of persons with a common interest to support chartering a non-scheduled flight.

a. An aircraft at Eglin AFB would be painted and numbered as an exact duplicate for a civil registered aircraft belonging to a CIA proprietary organization in the Miami area. At a designated time the duplicate would be substituted for the actual civil aircraft and would be loaded with the selected passengers, all boarded under carefully prepared aliases. The actual registered aircraft would be converted to a drone.

b. Take off times of the drone aircraft and the actual aircraft will be scheduled to allow a rendezvous south of Florida. From the rendezvous point the passenger-carrying aircraft will descend to minimum altitude and go directly into an auxiliary field at Eglin AFB where arrangements will have been made to evacuate the passengers and return the aircraft to its original status. The drone aircraft meanwhile will continue to fly the filed flight plan. When over Cuba the drone will being transmitting on the international distress frequency a "MAY DAY" message stating he is under attack by Cuban MIG aircraft. The transmission will be interrupted by destruction of the aircraft which will be triggered by radio signal. This will allow ICAO radio

Annex to Appendix
to Enclosure A

TOP SECRET — SPECIAL HANDLING — NOFORN

---

9. It is possible to create an incident which will make it appear that Communist Cuban MIGs have destroyed a USAF aircraft over international waters in an unprovoked attack.

   a. Approximately 4 or 5 F-101 aircraft will be dipatched in trail from Homestead AFB, Florida, to the vicinity of Cuba. Their mission will be to reverse course and simulate faking aircraft for an air defense exercise in southern Florida. Thses aircraft would conduct variations of these flights at frequent intervals. Crews would be briefed to reamin at least 12 miles off the Cuban coast; however, they would be required to carry live ammunition in the even that hostile actions were taken by the Cuban MIGs.

   b. On one such flight, a pre-briefed pilot would fly tail-end Charley at considerable interval between aircraft. While near the Cuban Island this pilot would broadcast that he had been jumped by MIGs and was going down. No other calls would be made. The pilot would then fly directly west at extreamly low altitude and land at a secure base, an Eglin auxiliary. The aircraft would be met by the proper peple, quickly stored and given a new tail number. The pilot who had performed the mission under an alias, would return to his normal place of business. The pilot and aircraft would then have disappeared.

   c. At precisely the same time that the aircraft was presumably shot down a submarine or small surface craft would disburse F-101 parts, parachute, etc., at approximately 15 to 20 miles off the Cuban coast and depart. The pilots returning to Homestead would have a true story as far as they knew. Search ships and aircraft coulsd be dispatched and parts of aircraft found.

Annex I

ENCLOSURE B

FACTS BEARING ON THE PROBLEM

1. The Joint Chiefs of Staff have previously stated that US unilateral military intervention in Cuba cab be undertaken in the event that the Cuban regime commits hostile acts against US forces or property which would serve as an incident upon which to base covert intervention.

2. The need for positive action in the event that current covert efforts to foster an internal Cuban rebellion are unsuccessful was indicated** by the Joint Chiefs of Staff on 7 March 1962, as follows:

"- - - determination that a credible internal revolt is impossible of attainment during the next 9-10 months will require a decision by the United States to develop a Cuban "provocation" as justification for positive US military action."

3. It is understood that the Department of State also is preparing suggested courses of action to develop justification for US military intervention in Cuba.

Document finishes

This horrific list of proposed JCS generated false-flags or self-acts of terrorism was intended to be spun to generate public and congressional support for a conventional military invasion of Cuba, an invasion that would be highly profitable for the MIC, and re-enforce the sentiment of Cold War Soviet malice.

However, several points regarding the *Operation Northwoods* briefing documents must be considered:

1) *Northwoods* was rejected out of hand by President John F. Kennedy, and as such did not go operational under his Presidency. Of course, Northwoods *was* fully supported by *every* member of the Joint Chiefs of Staff, and the necessary members of the Pentagon, and the CIA. This raises a nasty question: with Kennedy's assassination on November 22, 1963, was the Northwoods mentality immediately reinstated and continued under Lyndon Baines Johnson?

2) Defense Secretary Robert McNamara was forced by Kennedy to remove Lemnitzer from his position as Chairman of the Joint Chiefs of Staff; albeit to the highly sort-after post of U.S. Commander of NATO in Europe. Note that the document states that NATO must not be alerted to the planned illegal activity.

3) Lemnitzer knew that what he and the JCS had outlined in the Op Northwoods document was

illegal, because according to General B.W. Grey, [L]emnitzer feared Congressional investigation and ordered all evidence destroyed. General Grey's notes are the only record of what happened in the inner sanctum of the JCS in the early 1960s.

4) *Northwoods* was stopped at the final check by JFK - but the work of the Pentagon and the JCS in the twelve months prior shows a clear and obvious mentality to use deception to achieve military/ corporate goals - and the worst kind of deception at that - false-flag violence directed at *us*.

5) Had LBJ been in power, American boys may have been maimed and killed in a war triggered by a series of lies, and;

6) The false-flag incidents themselves would have maimed and killed American citizens on American streets.

Operation Northwoods did not happen for the Joint Chiefs of Staff in regards to Cuba and Castro, but did the Pentagon's mind for such deception continue to persist? Did elements of Northwoods evolve over the next twenty-four months into the trigger for another war - the war in Vietnam?

President Kennedy was quoted as wanting to 'draw a line in the sand' in reference to stopping the spread of communism, and his line was drawn through Vietnam. Some believe that immediately prior to his assassination in 1963, that Kennedy was set to de-escalate the Vietnam situation, which had 16,000 'military advisors' on the ground in country. It was said that Kennedy had suddenly 'gone soft on communism'. But it is of great interest to the false-flag researcher that with Kennedy's assassination and LBJ's appointment, Pentagon false-flag operations were immediately resumed - co-incidence?

The *Northwoods* document also makes reference to the 'drone-ization' of both surface ships, and civil aircraft, a significant advancement in the year of 1962 (but only in the public mind). In effect, *Northwoods* called for the drone-ization of aircraft to effect false acts of terrorism to justify the illegal invasion of an innocent nation.

The question for us is: has this mindset continued? Take a serious look at the triggering event prior to each war since Vietnam, and ask: is this a modern day *Operation Northwoods*?

~

www.ingramcontent.com/pod-product-compliance
Lightning Source LLC
Chambersburg PA
CBHW070856180526
45168CB00005B/1838